Chemical Warfare on America

To Heather, a great friend

George Orville

GEORGE ORVILLE

First Edition

PAGE PUBLISHING, INC.
New York, NY

First originally published by Page Publishing, Inc. 2018

This book is not intended to provide medical advice, diagnosis, or treatment. It is strictly for information only.

ISBN 978-1-64138-190-1 (Paperback)
ISBN 978-1-64138-191-8 (Digital)

Printed in the United States of America

This book is dedicated to
my mother, Katherine

Contents

Preface

I have to thank my mother, Katherine, who introduced me to natural solutions to multitude, non-life-threatening health problems. With the advent of Internet and subscribing to a number of doctors' emails regarding alternative health solutions, I became interested in natural, holistic medicine. On numerous occasions, the reports from most of the doctors using conventional and alternative medicine, the subject of fluoride and other toxins came up. Upon my retirement and having much free time on my hands, I decided to investigate the subject regarding fluoride safety and/or toxic effects on human body. To my astonishing surprise, I uncovered a wealth of information regarding fluoride's derogatory effects on human body. This led me also to investigate additional toxins present in our food and air. There are hundreds upon hundreds of scientific studies performed on these toxins exerting their toxic effect on our bodies.

To write about the adverse health effects on the human body, I asked myself the following question—would a reader understand the adverse health on a specific part of the body? I decided to at least give the reader some semblance of understanding in lay terms, as to what adverse effects are from a given toxin on our body. Therefore, I included a limited amount of descriptive discussions regarding our brain function, endocrinology (the gland system including the thyroid and adrenal), the nervous system, and the skeletal system. The reader is encouraged to access the numerous Internet links provided to further increase his/her understanding how our human body functions.

This book is divided into separate sections as to the human body and its function, such as endocrinology and how the glands function in our bodies. The discussions are devoted to the main major toxins, namely, fluoride in the water supply system, aspartame, MSG (monosodium glutamate), flavor enhancers in our food, chemtrails in our air, GMOs (genetically modified organisms) in our food, and pesticides, herbicides, insecticides, and vaccinations.

The reader is strongly encouraged to access all the Internet links provided for a much deeper discussion and understanding how these toxins affect our lives and how we are misled by our news media and our federal government agencies regarding our health.

To accomplish my task, I want to thank the following health organizations, medical doctors, and natural practitioners that have provided me with the information for my book. These are the following:

> Mike Adams, publisher of Natural News, whom one can subscribe to. Mike also participates on a Natural News radio program.
>
> Dr. Joseph Mercola, who provides daily health reports via emails that one can subscribe to.
>
> Dr. Russell Blaylock, neurosurgeon, who has written books and provides health tips via Internet.
>
> PEERS, who publishes WantToKnow, have a wealth of information on numerous subjects by compiling news articles from newspapers around the world.
>
> Health Science Institute, which has a number of health conscious doctors that provide conventional and alternative medical tips. HSI also publishes a book on alternative holistic cures titled *Cures from the Vault.*
>
> Harvard Medical School newsletter.
>
> Laissez Faire Club editor, G. Hill.
>
> Dr. E. Ben Joseph, a naturopathic doctor, who also has a weekly radio program on Saturday mornings from El Paso, Texas.

Even federal organizations like FDA (Federal Food and Drug Administration), NIH (National Institute of Health), CDC (Centers for Disease Control) can sometimes provide interesting admissions regarding our health.

Holmestead, Ontario, Canada, regarding chemtrails.

Dr. Michael Castle, an environmental professional.

Fluoride Action News.

Geoengineeringwatch.com.

There are numerous other health conscious organizations that can be accessed on the Internet. I am highly indebted to them for all for the information that these individuals and organizations have provided. Please understand that the information compiled in this book is just a very tip of the iceberg. There are numerous books written on the individual subjects that I am encouraging the reader to follow up on.

Some of these books are as follows:

Fluoride:
The Fluoride Deception by Christopher Bryson
The Case Against Fluoride by Paul Connett
The Hundred Year Lie by Randall Fitzgerald
The Devil's Poison by Dean Murphy

MSG (Monosodium glutamate):
MSG is Everywhere by Daniel A. Twogood
In Bad Taste: The MSG Syndrome by George R. Schwartby
Battling MSG Myth by Deborah L. Anglesey
The Slow Poisoning of America by T. Michelle Erb

Aspartame:
Excitotoxins: The Taste That Kills by Dr. Russell Blaylock
Health and Nutrition by Dr. Russell Blaylock

Sweet Poison by Dr. Jantet Starr Hill
Aspartame, MSG and Excitotoxins by Mike Adams
Aspartic Disease: An Ingrained Epidemic by H. J. Roberts
Aspartame, Is It Safe? by H. J. Roberts
Sweet Deception by Joseph Smith
Aspartame Side Effects by Ron Harper
Aspartame Physiology and Biochemistry by Fifer Stegink

GMOs:
Seeds of Deception by Jeffrey M. Smith
The Genetic Roulette by Jeffrey M. Smith
The Seeds of Destruction by F. William Engdahl
A Genetically Engineered Food by Ronnie Cummins
Plain Truth from Mad Cowboy by Howard F. Lyman

Chapter 1

Human Body

The human body is a finely tuned mechanism that relies on chemical components to maintain a constant state of balance. State of being in balance is called homeostasis. If one or more of these chemicals fall out of balance from either an increase or a decrease in their levels, this may cause the systems of the body to work less efficiently. Chemical imbalances can occur for a number of reasons and may affect the body in a variety of ways.

For centuries, man was supplied with all the necessary nutrients necessary for his nutritional chemical balance from natural agricultural products. For centuries, he was able to use natural means to combat some of the ailments that afflicted him. He was practicing what the Greek doctor Hippocrates advocated over two thousand years ago—"Let thy food be your medicine and thy medicine your food." Over the centuries, as more and more science came into prominence to analyze the diseases affecting man, this practice has gone by the wayside. Man learned that he could come up with medicines produced in a test tube. I remember reading in a Polish novel that the wounded soldiers would put molded bread onto their wounds. Little did they know that it contained penicillin. Once it was discovered by Alexander Fleming, the age of antibiotics was ushered in. Numerous antibiotics came into play, treating a number of bacterial diseases. The problem that developed over the years is an abusive overuse of these antibiotics, and the microbes, through mutations, developed a resistance to the antibiotics. Science has determined that the diseases affecting humanity is done by bacteria, viruses, and fungi. To date, there is very limited number of antivirus medications. A prime example is cancer. Numerous theories have come into play what causes the cancer virus.

To compound the problem, man decided to improve the life through the world of chemistry. Initially, it was determined to use chemical compounds in agriculture to diminish the destructive effect of insects on the crops. The use of pesticides DDT and DDE came into use. Thank goodness that a person by the name of Rachel Carson who wrote a book titled *A Silent Spring* in which she detailed pesticides' destructive effect on the environment and life in general. The use of chemical compounds and elements, such as chlorine, were also used in water purification. It was determined that adding chlorine was a highly effective way in destroying bacteria and other harmful organisms in water. In the 1950s, another chemical was added in the form of fluoride, acclaimed as a good agent to decrease dental cavities. To date no good scientific study has been established as to the effectiveness of this chemical, regarding dental cavities.

This book is dedicated to examine the effect of toxins on the human body, specifically the brain and the endocrine system. To get a good understanding of the adverse effects on the brain and endocrine systems, a brief description of the brain functions and each important glandular function is also provided.

Some of the more prominent chemical compounds used are the halogen elements widely used as cleaning agents, surfactants, insecticides, and herbicides, as well as in medicine and water purification. Chlorine and fluorine are added to drinking water supply plants. Since that time, fluoridation has been banned in Western Europe. Aspartame, a synthetic sweetener, is added to numerous food products, even though initial studies indicated strong adverse effects, which were hidden from its initial application to the FDA for approval. Then there is MSG (monosodium glutamate), another food additive that was grandfathered into our food supply system as a taste enhancer without any scientific studies to determine its safety or detrimental effect on the human body. There are other food additives, such as azodicarbonamide, that is banned in other countries but used in the US without proper scientific studies. Then there's advent of chemtrails and all their toxic content without public's approval. Then man decided to play God by genetically modifying our food crops. Again no lengthy scientific studies were performed to prove their safety, but instead reports are coming in about the adverse effects. Also, a report from India lists about three hundred farmers committing suicide due to their genetic (GMO) crop failure, driving them into bankruptcy. To this list one can add the numerous pesticides, herbicides, and insecticides that are poisoning our environment, as well as numerous toxic, artificial medicines in form of inoculations, tablets, and pills. It is a shame that the pharmacological companies look only to synthetic solutions and down play with a vengeance any natural solutions to human ailments.

Regarding pesticides, as an example, this is what was happening in 2014 in a town of St. Louis, Michigan—cleaning up DDT contamination. The town had Velsicol Chemical Corp. plant that was producing the pesticide DDT. For a number of decades, the town people were reporting birds falling out of the sky. The birds' sudden deaths were from feeding on contaminated worms, grubs, and insects, poisoned by the tainted soil. Forensic studies of the dead birds indicated brain and liver abnormalities. The bird's average DDT content from the collected bird brains was 552 parts per million of DDT; 30 parts per million are lethal to the birds. This is a number of decades after the plant was shut down in 1977. The EPA is involved in removing the contaminated soil anyplace from six inches to about four feet deep, depending on the test results. This amount is designed to remove the contamination causing an ecological and human health risks.

Folks, think about this. This is so many decades after it was diagnosed as being toxic to the environment. Since then, we have numerous additional pesticides, herbicides like Agent Orange, and insecticides in our environment. Another incident involving Velsicol Chemical Corp. was the disaster of accidentally contaminating cattle feed with polybrominated biphenyl (PBB), which is a fire retardant. Thousands of cattle and other livestock were poisoned. Number of farms were quarantined. People in Michigan were exposed to a chemical linked to cancer, reproductive problems, and endocrine disruption. Presently, the GMO food is used extensively in numerous food products, including in livestock feed. The manufacturer Monsanto has a standing order not to use GMO food in their restaurants. Do they hide something from us? GMO crops from US are refused by numerous countries worldwide. This is contrary to the statement by Monsanto about feeding the world. Nobody wants this poisoned stuff from us, but it is being used here in our country. I guess it is OK for Monsanto and the White House to eat non-GMO food, but it is OK to feed the rest of us this poisoned food. GMO feed is also fed to our livestock. No concrete studies have been done as to the safety of the GMO food crops, but some studies indicate adverse effects on our health. Initially the adverse effect reported is that our honeybees, critical to pollination of numerous food crops, are being decimated due to the wide use of pesticides, herbicides, and insecticides.

To get a better understanding how all these toxins and poisons affect the function of the brain, the endocrine system, and the glands, we have to discuss the toxins in general terms and the effect they have on the human body. The first toxin to be discussed and its effect on our body is the toxin fluoride since it is added to our water plants throughout the country and because of the effect it has physiologically and then psychologically (autism, ADHD, ADD, etc.).

Function of the Human Brain

Think of the human brain as a central computer system that regulates a number of critical functions in the human body through a nervous system. It sends signals through the nerves to all parts of the body to make it work. As an example—heartbeat. The nervous systems control every part of our daily life, from breathing to memorizing facts. Sensory nerves gather information from the environment and send it to the spinal cord, which speeds the message to the brain. The brain is the control system of the central nervous system (CNS), which consists of the brain and the peripheral nervous system (PNS) made up of nerve cells outside the central nervous system, located throughout the body. The brain consists of three parts—hindbrain, which contains the cerebellum, pons, and medulla, sometimes referred to as the brainstem; midbrain, which contains the tectum (contains visual and auditory receptacles) and tegmentum (controls motor functioning and regulates awareness, attention, and some autonomic functions); and forebrain, which contains the cerebrum, thalamus, and hypothalamus.

Cerebellum, also known as the cortex, is the largest part of the human brain, associated with higher brain functions such as thought and action. It is divided into four sections or lobes. The four lobes are the frontal, parietal, occipital, and temporal:

1) Frontal lobe—associated with reasoning, planning, parts of speech, movement, emotions, and problem solving
2) Parietal lobe—associated with movement, orientation, recognition, perception, and stimuli
3) Occipital lobe—associated with visual processing
4) Temporal lobe—associated with perception and recognition of auditory stimuli, memory, and speech

The cerebral cortex is highly wrinkled, increasing the surface area and the amount of neurons (nerve cells) in it. The cerebrum is divided into two halves, known as the left and right hemisphere. They look to be symmetrical, but each side functions slightly differently. A bundle of axons (a long branching structure that is unique to a nerve cell) connects these two hemispheres. The surface of the cerebellum is gray, while the nerves underneath are white. The white nerve fibers carry signals between the nerve cells of the brain and other parts of the body. The cerebellum is similar to the cerebrum in that it also has two hemispheres. It is associated with regulation of movement, posture, and balance. Buried within the cerebrum is the limbic system, associated with the emotion. This system contains the thalamus, hypothalamus, amygdala, and hippocampus.

Thalamus is a large mass of gray matter situated in the forebrain, and it has sensory and motor functions. Hypothalamus is involved in functions, including homeostasis, emotion, thirst, hunger, and sleep-wake cycle, and controls autonomic nervous system. In addition, it controls the master gland, the pituitary. Amygdala located in the temporal lobe is involved with memory, emotion, and fear. Hippocampus is important for learning and memory, converting short-term memory to long-term memory, and recalling spatial relationship in the world around us. Brain stem is responsible for basic vital functions, such as breathing, heartbeat, and blood pressure. It is made up of midbrain, pons, and medulla.

Midbrain is that part of the brain, which includes tectum and tegmentum, which are involved in functions such as vision, hearing, eye movement, and body movement. Pons is involved in motor control and sensory analysis. It has parts that are important for the level of consciousness and for deep sleep. Medulla oblongata is responsible for maintaining vital body functions such as breathing and heartbeat. What a wonderful system our creator designed. All these human functions work together at the same time. Just imagine coordinating our movements, memory, and impulses from our environmental surroundings at the same time without missing a beat.

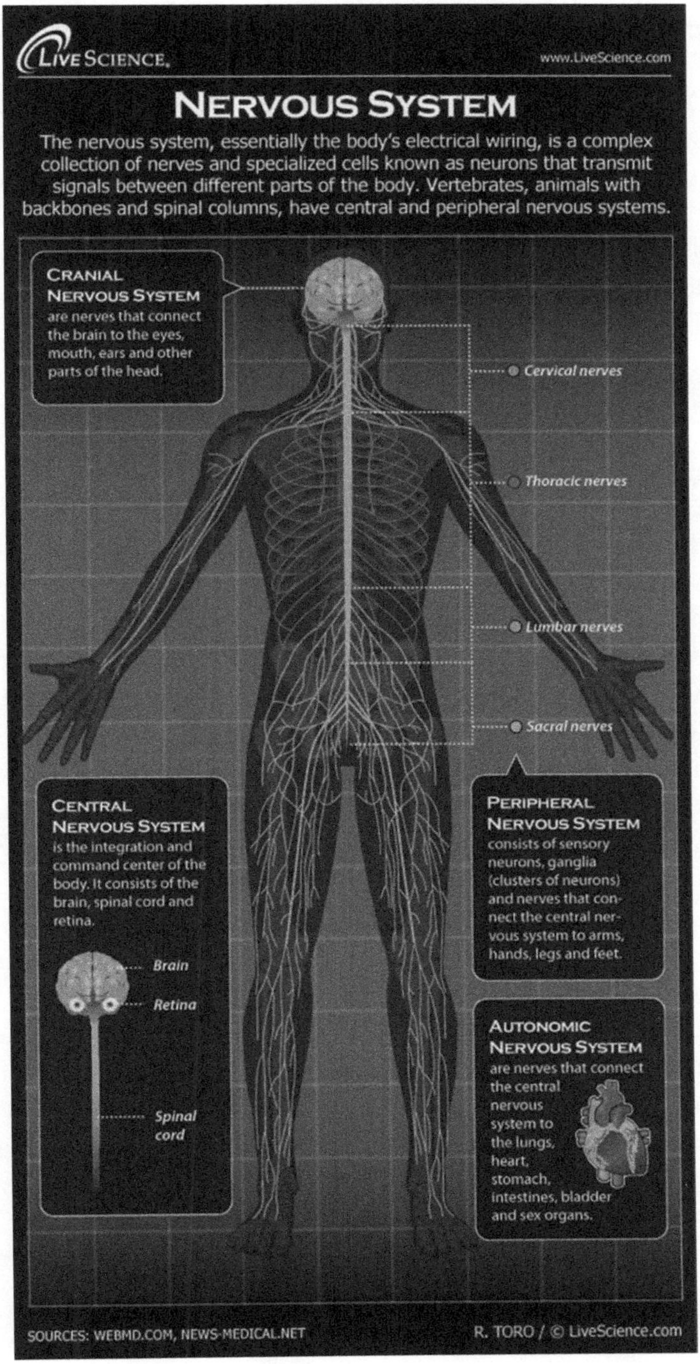
LIVE SCIENCE
www.LiveScience.com
NERVOUS SYSTEM
The nervous system, essentially the body's electrical wiring, is a complex collection of nerves and specialized cells known as neurons that transmit signals between different parts of the body. Vertebrates, animals with backbones and spinal columns, have central and peripheral nervous systems.
CRANIAL NERVOUS SYSTEM are nerves that connect the brain to the eyes, mouth, ears and other parts of the head.
Cervical nerves
Thoracic nerves
Lumbar nerves
Sacral nerves
CENTRAL NERVOUS SYSTEM is the integration and command center of the body. It consists of the brain, spinal cord and retina.
Brain
Retina
Spinal cord
PERIPHERAL NERVOUS SYSTEM consists of sensory neurons, ganglia (clusters of neurons) and nerves that connect the central nervous system to arms, hands, legs and feet.
AUTONOMIC NERVOUS SYSTEM are nerves that connect the central nervous system to the lungs, heart, stomach, intestines, bladder and sex organs.
SOURCES: WEBMD.COM, NEWS-MEDICAL.NET
R. TORO / © LiveScience.com

Nervous system function

Neurons in the brain communicate via electrical signals and neurotransmitters. The nervous system is a complex collection of nerves and specialized cells known as neurons that transmit signals between different parts of the body. Animals with backbones and spinal columns have central and peripheral nervous systems. The central nervous system is made up of the brain, spinal cord, and retina. The peripheral nervous system consists of sensory neurons, ganglia (clusters of neurons), and nerves that connect to one another and to the central nervous system.

Description of the nervous system

The nervous system is essentially the body's electrical wiring. It is composed of nerves, which are bundles of fibers that start at the brain and central cord and branch out to every other part of the body. Neurons send signals to other cells through thin fibers called axons, which cause chemicals known as neurotransmitters to be released at junctions called synapses. A synapse gives a command to the cell and the entire communication process typically takes only a fraction of a millisecond. Sensory neurons react to physical stimuli, such as light, sound, and touch, and send feedback to the central nervous system about the body's surrounding environment. Motor neurons, located in the central nervous system or in peripheral ganglia, transmit signals to activate the muscles or glands. Glial cells, derived from the Greek word for "glue," support the neurons and hold them in place. Glial cells also feed nutrients to neurons, destroy pathogens, remove dead neurons, and act as traffic cop by directing the axons of neurons to their targets. Specific types of glial cells (oligodendrocytes in the central nervous system and Schwann cells in the peripheral nervous system) generate layers of a fatty substance called myelin, which wraps around axons and provides electrical insulation to enable them to rapidly and efficiently transmit signals.

Diseases of the nervous system

There are a number of tests and procedures to diagnose conditions involving the nervous system. Aside from MRIs and CT scans, an electroencephalogram (EEG) is often used to record the brain's continuous electrical activity by attaching electrodes to the scalp. Positron emission tomography (PET) is a procedure that measures the metabolic activity of cells. A spinal tap places a needle into the spinal canal to drain a small amount of cerebral spinal fluid that is tested for infection or other abnormalities. A number of nerve disorders can affect the nervous system, including vascular disorders such as the following:

- stroke
- transient ischemic attack (TIA)
- subarachnoid hemorrhage
- subdural hemorrhage and hematoma
- extradural hemorrhage

The nervous system can also experience functional difficulties, which result in conditions such as the following:

- epilepsy
- Parkinson's disease
- multiple sclerosis
- amyotrophic lateral sclerosis (ALS), also known as Lou Gehrig's disease
- Huntington's chorea
- Alzheimer's disease

Infections, such as meningitis, encephalitis, polio, and epidural abscess, can also affect the nervous system. Structural disorders, such as brain or spinal cord injury, Bell's palsy, cervical spondylosis, carpal tunnel syndrome, brain or spinal cord tumors, peripheral neuropathy, and Guillain-Barré syndrome, also strike the nervous system.

Study of the nervous system

The branch of medicine that studies and treats the nervous system is called neurology, and doctors who practice in this field of medicine are called neurologists. Neurosurgeons perform surgeries involving the nervous system. There are also psychiatrists and psychologists, who specialize in behavioral and rehabilitation on patients who have experienced a disease or injury to their brain or nervous system that impacts their ability to perform. http://www.lifescript.com/Health

Toxins and Poisons that Affect the Brain

Toxins

There are two forms of toxins. One is produced by a synthetic process and the second produced by living organisms—both poisonous to human beings. Toxins can be small molecules, peptides, or proteins capable causing diseases on contact or absorption by body tissues. Toxins vary greatly in their effect, ranging from usually minor to acute as in bee stings, or deadly as in botulism (botulinum toxin). Toxin implies that it is a biological poison, a product of plants, animals, and microorganism, such as bacteria, viruses, fungi, ricketsiae, or protozoa. http://www,en,wikipedia.org/wiki/Toxin

Poisons

Poisons are substances that cause disturbances to organisms by chemical reactions or other activity on the molecular scale. Poisons are most often applied in industry, agriculture, and other uses specifically applied for their toxic effect—for example, pesticides, herbicides, insecticides, and water purification by adding chlorine to kill the microorganisms. Poisons, throughout human history, have been intentionally used as a method of murder, suicide, execution, and pest-control. Acute poisoning is exposure to a poison for short periods of time. Symptoms may occur immediately or within a very short time, depending on the concentration of the poison at the time of inhalation or absorption through a body tissue. Chronic poisoning is a long-term, repeated, or continuous exposure to a poison, where symptoms do not occur immediately, but after some extended period of time, such as months or years. Higher concentrations of a poison will exhibit symptoms sooner, while lower concentrations will take longer for any symptoms to appear. http://www.en.wikipedia.org/wiki/poisons

Toxicology is the study of the symptoms, mechanisms, treatment, and diagnosis of biological poisoning. Most biocides, such as pesticides, insecticides, and herbicides, are created poisons to target organisms, insects, pests, and plants. Chronic poisoning can occur in nontargeted organisms, such as humans, and beneficial organisms, such as bees. The following discussion is about manmade poisons affecting the nervous system in humans, as well as the endocrinological systems (gland system) in human beings.

Toxic effects affecting the neurotransmission of neurons (nerve cells)

Certain so-called neurotoxins do not alter the nerve cells structure but interfere with the transmission of impulses of the neuron cell. The most common neurotransmitters are acetylcholine and norepinephrin, but are complimented by other numerous amine neurotransmitters and amino acids. Neurotransmitters are chemical messengers that carry signals between neurons and other cells of the body.

Neuropathy

Neuropathy is any disease of the nervous system. It can be caused by a family of drugs called barbiturates. These drugs induce anoxia (a condition characterized by an absence of oxygen supply to an organ or tissue), in that instance to the brain. Cyanide and azide inhibit cytochrome-c oxidase, resulting in cytotoxic anoxia. Cytochrome-c oxidase is a membrane protein that controls the last step of food oxidation. The neuron cell body is also directly affected by ethylmercury (used in inoculations), which destroys the neuron. Doxorubicin is an anticancer chemotherapy drug; it damages and interferes with DNA and inhibits the synthesis of RNA. DNA is a substance carrying organism's genetic information. It is a nucleic acid molecule in the form of a twisted double-strand helix that is the major component of the chromosome. It carries the hereditary material in humans. RNA molecules are single-stranded nucleic acids composed of nucleotides. It plays an important role in protein synthesis as it is involved in the transcription, decoding, and translation of the genetic code to produce protein. It is also called ribonucleic acid.

Organic toxins used as pesticide accumulate in the Golgi structures, which result in cell swelling and neurosis. The Golgi apparatus is a series of membranes shaped like pancakes. It is the distribution and delivery of the cell's chemical products. It assists in the modification of proteins and other macromolecules. Neurosis is a functional disorder typified by excessive anxiety or indecision and a degree of social maladjustment. Glutamate affects the dendrites and has neuroexcitatory effect. Alcohol in pregnant women can result in abnormal neuronal migration in their offspring with occurrence of dendrite spines. Dendrite is any of the short-branched threadlike extensions of a nerve cell, which conduct impulses toward the cell body. Dendrite spine is a small membranous protrusion from a neuron's dendrite that typically receives input from a single synapse of an axon. Interesting to note, that women during pregnancy are warned not to consume alcohol. How about warning women not to consume food containing MSG (monosodium glutamate) during pregnancy? The glutamate's effect of neuroexcitatory effect means that the neuron gets too excited to the point of destruction. Hexacarbons (n-hexane and n-butylketone) result in an axon swelling. Axon is a long slender projection of a nerve cell or neuron, typically conducting electrical impulses away from the neuron's cell body. Clioquinol is an antifungal drug and antiprotozoan drug; it is a neurotoxin in large doses, inhibiting certain enzyme related to DNA replication.

Interference with impulse conduction

These poisons target mainly the nerve membranes. Tetrodotoxin and saxitoxin block the neuronal sodium channels that can lead to death due to respiratory failure. DDT and pyrethroids have the same effect, interference with synaptic transmission. Botulinum toxin causes paralysis of muscles due to impairment of neurotransmitter acetylcholine from the motor cell nerve endings. Black widow spider venom does the opposite, but the result is the same—paralysis. Tetanospasmin, a neurotoxin component of the exotoxin tetanus toxin, blocks the release of inhibitory amino acid transmitters causing spastic paralysis by reaching the central nervous system by retrograde axonal transport. Spastic means lacking physical coordination.

This is a brief summary of some of the brain terminology, what their function is, and how some toxins and poisons are affecting our nervous systems, both central and peripheral. More concrete discussions of the toxic effects of fluoride, aspartame, MSG, and GMOs (genetically modified organisms in food crops) will be presented under each individual subject matter. This also includes glyphosate, a herbicide extensively used by the agricultural industry as well as consumers in weed control.

A Silent Epidemic of Neurotoxins Threatening Children's Brains

One of the better summaries regarding neurotoxicity is the following article by Kaherine Martinko. It is reprinted in its entirety from http://www.treehugger.com/silent-epidemic-neurotoxins-threatening . . .

Katherine Martinko (@feistyredhair), Living / Health, March 6, 2014

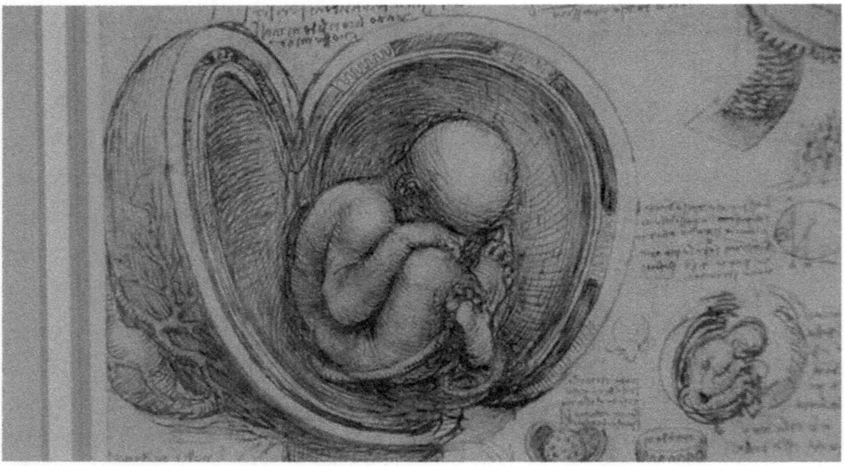

Leonardo da Vinci's drawing of a fetus

A plethora of toxic chemicals is endangering and poisoning our children's brains. The problem is so serious that two American doctors are calling for a global overhaul of the chemical regulatory process in order to protect children's brains, but regulations cannot—or will not, due to industry pressure—keep up with the research that continues to reveal neurotoxins all around us. The biggest window of vulnerability to chemicals is between conception and early childhood. This is serious because any negative effects on the brain are permanent. "When this happens in children or during pregnancy, those chemicals are extremely toxic, because we now know that the developing brain is a uniquely vulnerable organ.

Dr. Philip Landrigan of Mount Sinai School of Medicine in New York and Dr. Philippe Grandjean of the Harvard School of Public Health in Boston have spent 30 years studying industrial chemicals and have published lists of the worst neurotoxins. These "impact brain development and can cause a number of neurodevelopmental disabilities including attention-deficit hyperactivity disorder, autism, dyslexia and other cognitive damage.

The neurotoxins that the doctors have identified so far are lead, methylmercury, arsenic, polychlorinated biphenyls (or PCBs, which have been banned in the US since 1979), toluene (used in household products like paint thinner, nail polish, spot remover), manganese, fluoride, tetrachloroethylene (a solvent), polybrominated diphenyl ethers (flame retardants), and two agricultural pesticides, one of which is DDT. Cosmetics, which contain phthalates in the US but are banned in Europe, are another major area of concern.

Fluoride might come as a surprise to those who live in towns or cities where tap water contains fluoride. Landrigan and Grandjean looked at 27 studies of children in China who were exposed to high levels of fluoride in drinking water. The data suggests a decline of about seven IQ points on average among those children. Landrigan and Grandjean believe that all new chemicals and those currently in use must be tested for developmental neurotoxicity. "We have the test methods and protocols to determine if chemicals are toxic to brain cells [but] it's a matter of political will. We have tried in this country over the last decade to pass chemical safety legislation but the chemical industry and their supporters have successfully beat back the effort.

As a parent, I realize I can't wait around for the chemical industry to clean up its act, so that's why it's crucial to take personal steps toward detoxifying my children's lives. My strategy includes buying second-hand clothes that won't off-gas like new ones do; choosing toys made of natural materials over plastic ones; using all-natural bath and body care products, such as pure castile soap and coconut oil; using natural household cleaners; avoiding the use of neurotoxin-laden cosmetics for myself; and minimizing the presence of plastic in the house.

Chapter 2

Fluorine

Numerous studies have been done on the harmful effects of fluoride addition to water in the form of fluoride salt at water treatment plants. Potential for disaster exists as to how much should be added. Disaster like that occurred in Alaska. Most surprising is that the harmful effects of fluoride have been known by conventional medical organizations for over half a century. For example, the *Journal of the American Medical Association* (JAMA) stated in their September 18, 1943, issue that **"fluorides are general protoplasmic poisons that change the permeability of the cell membranes by certain enzymes."** An editorial published in the *Journal of the American Dental Association*, October 1, 1944, stated, **"Drinking water containing as little as 1.2 ppm (parts per million) fluoride will cause developmental disturbances. We cannot run the risk of producing such serious systemic disturbances. The potentialities for harm outweigh the good."**

Long lost research linking fluoride to cancer has resurfaced in a Dutch film clip featuring Dr. Dean Burk, who in 1937, cofounded the US National Cancer Institute (NCI) and was in charge of its cytochemistry department for over thirty years. In the taped interview, he equates water fluoridation to **"PUBLIC MURDER,"** referring to a study that had been done on the ten largest US cities with fluoridation compared to the ten largest cities without it. The study clearly demonstrated that deaths from cancer abruptly rose in as little as a year or two after fluoridation began. This and other studies linking fluoride to cancer were government ordered but were quickly buried once fluoride was found to be linked to dramatic increases in cancer. The only real solution is to stop the archaic practice of water fluoridation in the first place.

Part of the problem is that it's an accumulative toxin that, over time, can lead to significant health problems that are not immediately linked to fluoride overexposure. In a 2005 paper entitled "Fluoride—A Modern Waste" by Lita Lee, PhD, states, "Yiamouyiannis' book 'Fluoride: The Aging Factor' documents the cumulative effect of tissue damage by fluoride, commonly seen as aging (collagen damage), skin rashes and acne, gastrointestinal disorders, and osteoporosis, to name a few. The US Centers for Disease Control (CDC) and the Safe Water Foundation, reported that 30,000 to 50,000 excess deaths occur in the US each year in areas in which the H_2O (water) contains 1 ppm of fluoride."

There's no doubt about it—fluoride should not be ingested. Even scientists from EPA's (Environmental Protection Agency) National Health and Environmental Effects Research Laboratory have classified fluoride as a **"chemical having substantial evidence of developmental neurotoxicity."** Furthermore, according to the Centers for Disease and Control (CDC), 41% of American adolescents now have dental fluorosis—unattractive discoloration and mottling of the teeth that indicate overexposure to fluoride. Clearly, children are being overexposed and their health and development put in jeopardy.

Ten facts about fluoride that one needs to know:

1) Most developed countries do not fluoridate their water. More people drink fluoridated water in the US alone than the rest of the world combined. In Western Europe 97% of the population drink nonfluoridated water.
2) Fluoridated countries *do not* have less tooth decay than the nonfluoridated countries. According to the World Health Organization (WHO), there is *no discernible difference* in tooth decay between developed countries that fluoridate

their water and those that do not. The decline in tooth decay experienced in the US over the past sixty years, which is often attributed to fluoridated water, has likewise occurred in all developed countries, which do not fluoridate their water. It probably is due to better dental hygiene.

3) Fluoride affects many tissues in one's body beside teeth. The American Dental Association (ADA) assumes that consuming fluoride is only an issue that involves dental health. According to numerous studies, fluoride is an endocrine disruptor that affects your bones (skeletal fluorosis resulting in brittle bones), brain (autism, ADD, ADHD, muscular coordination), thyroid gland (hypothyroidism disrupting numerous metabolic functions, cause for obesity [?]), pineal gland, and even one's sugar (diabetes?). There have been thirty-four human studies and one hundred animal studies linking fluoride to brain damage, including lower IQ in children.

4) Fluoridation is not a natural process. Fluoride is naturally occurring in some areas, leading to high levels in certain water supplies naturally. Fluoridation advocates often use this to support its safety, but naturally occurring substances are not automatically safe. (How about uranium or arsenic as examples.) The fluoride added to water to most water supplies is not the naturally occurring substance, but rather fluorosilicic acid, which is captured in air pollution control devices of the phosphate fertilizer industry. This captured fluoride acid is the most contaminated chemical added to public water supplies and may impose additional risks to public health, in addition to those present by natural fluoride. These health risks include cancer hazard, from acid's elevated arsenic content (already proven in the 1939 study by the comparison of the ten largest cities that were on fluoridated water supply against ten largest cities that were not on fluoridated water supply system). There's also additional neurotoxic

hazard from the acid's ability to increase the erosion of lead from old pipes.

5) American teenagers, about 40%, already show visible signs of fluoride exposure. This sign is the dental fluorosis, a condition that refers to change in the appearance of tooth enamel that is caused by long-term fluoride ingestion during the time that the teeth are being formed. In some areas, fluorosis is as high as 70–80%, and some children are suffering from even advanced form of fluorosis. This is more than likely that this is a sign that these children are receiving large amounts of fluoride from multiple sources—fluoride from drinking water, fluoride from toothpaste, fluoridated mouthwashes, processed beverages/foods, and fluoridated pesticides. Not only do we need to address the issue of water fluoridation but how this exposure is magnified by other sources of fluoride that are now common. It is important to realize that dental fluorosis is not just cosmetic but is an indication that the rest of the body, such as your bones, muscles, and internal organs, including your brain, have been overexposed to fluoride poisoning. The visual detrimental effect on the surface of your teeth virtually guarantees that it's also damaging other parts of your body, such as bones, thyroid gland, brain, and adrenal gland, creating a multitude of other health risks.

6) Fluoridated water for infants provides no benefits but numerous health risks. Infants who consume formula made from fluoridated tap water may consume up to 1,200 micrograms of fluoride or about one hundred times more than the recommended safe amounts. Such spikes of fluoride exposure during infancy provide no known advantage to the teeth, but they do have plenty of known harmful effects. Babies given fluoridated water in their formula are not only more likely to develop dental fluorosis but will also have reduced IQ scores and other harmful health effects. A number of prominent dental researchers

now advise that parents should not add fluoridated water to baby formulas.

7) Fluoridated supplements have never been approved by the Federal Drug Administration (FDA).

8) Fluoride is added to drinking water to prevent a disease—tooth decay—but no such study exists to confirm this theory. As such, fluoride becomes a medicine by FDA definition. Proponents claim this is no different than adding vitamin D to milk, but fluoride *is not an essential nutrient.* Many European countries have rejected fluoride for the very reason that delivering medication via water supply would be inappropriate. Water fluoridation is a form of mass medication that denies one the right to informed consent.

9) Swallowing fluoride provides little if any benefit to teeth. It is now widely recognized that fluoride's only justifiable benefit comes from *topical* contact with teeth, which even the US, Centers for Disease and Control Prevention (CDC), has acknowledged. Adding it to water and/or swallowing pills offer no benefit to your teeth.

10) Disadvantaged communities are the most disadvantaged by fluoridating drinking water. Fluoride toxicity is magnified by conditions that occur much more frequently in low-income areas. This includes the following:

*Nutrient deficiency
*Infant formula consumption
*Kidney disease
*Diabetes

African American and Mexican American children have significantly higher rates of dental fluorosis, and many low-income urban communities also have severe oral health crises, despite decades of water fluoridation.
http://www.medicalnewstoday.com/article/154164, http://www.webmed.com/fluoride

One can also access following topics on fluoride: fluoride free water, latest fluoride news, fluoride poisoning (adverse health effects), fluoride (the invisible killer), infants overdosed with fluoride, the fluoride glut, Europe says *no* to fluoridated water, the dangers associated with fluoride. There are hundreds of sources to read about the dangers of the toxic pollutant fluoride and why EPA headquarters professional union opposes fluoridation. Fluoride endangers the thyroid. Government reports say fluoride added to water supply may harm the population.

Program F

The following article was reprinted in its entirety since this sheds important light how back in history fluoridation was being scrutinized. This article is quite extremely interesting since it goes back all the way to the times of the Manhattan Project (1940s).

Program F was during atomic energy study of fluoride effect on human health. The article is by Josh Griffiths and Chris Bryson:

Fluoride, teeth, and the atomic bomb

Some fifty years after the United States began adding fluoride to public drinking water supplies to reduce cavities in children's teeth, declassified government documents are shedding new light on the roots of that still—controversial public health measure, a surprising connection between fluoride and the drawing of nuclear age.

Today two thirds of US public drinking water is fluoridated. Many municipalities still resist the practice, disbelieving the government assurances of safety.

Since the days of World War II, when this nation prevailed by building the world's first atomic bomb, US public health leaders have maintained that the low doses of fluoride are safe for people, and good for children's teeth. This safety verdict should now be re-examined in the light of hundreds of once-secret WWII documents obtained by Griffiths and Bryson—including declassified papers of the Manhattan Project, the US military group that built the atomic bomb.

Fluoride was the key chemical in atomic bomb production according to the documents. Massive quantities of fluoride—millions of tons—were essential for the manufacture of bomb grade uranium and plutonium for nuclear weapons throughout the Cold War. One of the most toxic chemicals known, fluoride rapidly emerged as the leading chemical health hazard of the US atomic bomb program, both for workers and for nearby communities, the documents reveal.

Other revelations include:

Much of the original proof of fluoride is safe for humans in low doses was generated by A-bomb program scientists who had been secretly ordered to provide "evidence useful in litigation" against defense contractors for fluoride injury to citizens. The first lawsuit against the US A-bomb program were not over radiation, but over fluoride damage, the documents show. Human studies were required. Bomb program researchers played a leading role in the design and implementation of the most extensive US study of the health effects of fluoridating public drinking water—conducted in Newburgh, New York, from 1945 to 1956. Then, in a classified operation code-name "Program F," they secretly gathered and analyzed blood and tissue samples from Newburgh citizens with the cooperation of State Health Department personnel. **The original secret version—obtained by these reporters—of a 1948 study published by Program F scientists in the Journal of the American Dental Association shows that evidence of adverse health effects from fluoride was censored by the US Atomic Energy Commission (AEC), considered the most powerful of Cold War agencies, for reasons of national security.**

The bomb program fluoride safety studies were conducted at the University of Rochester, site of the most notorious human radiation experiments of the Cold War, in which unsuspecting hospital patients were injected with toxic doses of radioactive plutonium. The fluoride studies were conducted with the same ethical mindset in which "national security" was paramount. The US government conflict of interest—and its motive to prove fluoride "safe"—has not until now been made clear to the general public in the furious debate over water fluoridation since the 1950's, nor to civilian researchers and health professionals of journalists.

The declassified documents resonate with growing body of scientific evidence, and a growing chorus of questions, about the health effects of fluoride in the environment. Human exposure to fluoride has mushroomed since World War II, due not only to fluoridated water and toothpaste, but to environmental pollution by major industries from aluminum to pesticides, fluoride is a critical industrial chemical. The impact can be seen, literally, in the smiles of our children. Large number of US young people, up to 80% in some cities—now have dental fluorosis, the first visible signs of excessive fluoride exposure, according to the US National Research Council (NRC). The signs are whitish flecks or spots, particularly on the front teeth, or dark spots or stripes in more severe cases.

Now researchers, who have reviewed these declassified documents fear that Cold War national security consideration may have prevented objective scientific evaluation of vital public heath questions concerning fluoride.

"Information was buried," concludes Dr. Phyllis Mullenix, former head of toxicology at Forsyth Dental Center in the early 1990s indicated that fluoride was a powerful central nervous system (CNS) toxin, and might adversely affect human brain functioning, even at low doses.

New epidemiological evidence and physical brain cell damage observed under electron microscope, evidenced by the Yu, Dong, and Han studies from China, support the correlation between low-dose fluoride exposure of brain damage. Recent Harvard study indicates diminished IQ in children. Is this the smoking gun pointing to be the cause of autism, ADD, ADHD? Probably yes!)

Mullenix's results were published in 1995 in a reputable peer-reviewed scientific journal. During her investigation. Mullenix was astonished to discover there had been virtually no previous US studies of fluoride's effect on the human brain. Then, her application for a grant to continue her CNS research was turned down by the US National Institute of Health (NIH), where NIH panel, she says, finally told her that "fluoride does not have central nervous system effect."

What a blatant lie. Are they suppressing an ulterior motive?

Declassified documents of the US atomic-bomb program indicate otherwise. An April 29, 1944 Manhattan Project memo reports "Clinical evidence suggests that uranium hexafluoride may have a rather marked central nervous effect . . . it seems most likely that the F (code for fluoride) component rather than the T (code for uranium) is the causative factor.

The memo stamped "secret" is addressed to the head of the Manhattan Project's Medical Section, Colonel Stafford Warren. Colonel Warren is asked to approve a program of animal research on CNS effects. "Since work with these compounds is essential, it will be necessary to know in advance what mental effects may occur after exposure. This is important not only to protect a given individual, but also to prevent a confused workman from injuring others by improperly performing his duties." On the same day Colonel Warren approved the CNS research program. This was in 1944, at the height of the Second World War, and the nation's race to build the world's first atomic bomb. For research on fluoride's CNS effects to be approved at such a momentous time, the supportive evidence set forth in a proposal forwarded along with the memo, must have been persuasive. The proposal, however, is missing from the files of the US National Archives. "If you find the memo, but the document they refer to is missing, it probably is still classified." said Charles Reeves, chief librarian at the Atlanta branch of the US National Archives and Records Administration, where the memos were found. Similarly, no results of the Manhattan Project's fluoride CNS research could be found in the files. After reviewing the memos, Mullenix declared herself "flabbergasted." She went on "How could I be told by NIH that fluoride has no central nervous system effects when these documents were sitting there all the time?" She reason's that the Manhattan Project did do fluoride CNS studies in "that kind of warning, that fluoride workers might be a danger to the bomb program by improperly performing their duties—I can't imagine that would be ignored"—but that the results were buried because they might create a difficult legal and public relations problem to the government.

The author of the 1944 CNS research proposal was Dr. Harold C. Hodge at the time of fluoride toxicology studies for the University of Rochester division of the Manhattan Project. Nearly fifty years later at the Forsyth Dental Center in Boston, Dr. Mullenix was introduced to a gently ambling elderly man brought in to serve as a consultant on her CNS research—Harold C. Hodge. By that time Hodge had achieved status emeritus as a world authority of fluoride safety. "But even though he was supposed to be helping me,' says Mullenix, "he never once mentioned CNS work he had done for the Manhattan Project."

The "black hole" in fluoride CNS research since the days of the Manhattan Project is unacceptable to Mullenix, who refuses to abandon the issue. "There is so much fluoride exposure now, and we simply do not know what it is doing." she says, "You can't just way walk away from this."

Dr. Antonio Noronha, an NIH scientific review advisor familiar with Dr. Mullenix grant request, says her proposal was rejected by peer-review group. He terms her claim of institutional bias against CNS research "farfetched," he adds "We strive very hard at NIH to make sure politics does not enter the picture."

Doesn't this sound like fluoride infected liberal trying to divert the subject matter? He is the one playing politics and doing a government cover-up.

Fluoride and national security

The documentary trail begins at the height of WW2 in 1944, when a severe pollution incident occurred of the E. I du Pont du Nemours Company chemical factory in Deepwater, New Jersey. The factory was then producing millions of pounds of fluoride for the Manhattan Project, the ultra-secret US military program racing to produce the world's first atomic bomb. The farms downwind Gloucester and Salem counties were famous for their high quality produce. Their peaches went directly to the Waldorf Astoria Hotel in New York. Their tomatoes were bought up by Campbell's Soup. But in the summer of 1943, the farmers began to report that their crops were blighted, and that "something is burning up the peach crops around here." Poultry died after an all-night thunderstorm, they reported. Farm workers who ate the produce they had picked sometimes vomited all night and into the next day. "I remember our horses looked sick and were too stiff to work," these reporters were told by Michael Giordano, who was a teenager at the time. Since cows were so crippled they could not stand up, and gazed by crawling on their bellies. The account was confirmed by taped interviews, shortly before he died, by Philip Sadtler of Sadtler Laboratories of Philadelphia, one of the nation's oldest chemical firms. Sadtler had previously personally conducted the initial investigation of the damage. Although the farmers did not know it, the attention of the Manhattan Project and the federal government was riveted on the New Jersey incident, according to once-secret documents obtained by these reporters. After the war's end, in a secret Manhattan Project memo dated March 1, 1946, the Project's chief of fluoride toxicology studies, Harold C. Hodge, worriedly wrote to his boss Colonel Stafford L. Warren, Chief of the Medical Division, about "problems associated with the question of fluoride contamination of the atmosphere in a certain section of New Jersey. "There seems to be four distinct (though related) problems."

Hodge continues:

1) A question of injury to the peach crop in 1944
2) A report of extraordinary fluoride content of vegetables grown in the area
3) A report of abnormally high fluoride content in blood of human individuals residing in this area
4) A report raising the question of serious poisoning of horses and cattle in this area

The New Jersey farmers waited until the war was over, then sued du Pont and the Manhattan Project for fluoride damage—reportedly the first lawsuits against the US A-bomb program. Although, seemingly trivial, the lawsuits shook the government, the secret documents were convened in Washington, with compulsory attendance by scores of scientists and officials from the US War Department, the Manhattan Project, the Food and Drug Administration, the Agriculture and Justice Departments, the US Army's Chemical Warfare Service and Edgewood Arsenal, the Bureau of Standards, and du Pont lawyers. Declassified memos of the meetings reveal a secret mobilization of the full forces of the government to defeat the New Jersey farmers. These "agencies are making scientific investigations to obtain evidence which may be used to protect the interest of the Government at the trial of suits brought by owners of peach orchards in . . . New Jersey," stated Manhattan Project Lieutenant Colonel Cooper B. Rhodes, in a memo cc'd to General Groves.

> *August 25, 1845*
> *Subject investigation of Crop Damage at Lower Penns Neck, New Jersey.*
> *To The Commanding General Army Services, Forces Pentagon Building, Washington, D.C.*
>
> *At the request of the Secretary of War, the Department of Agriculture has agreed to cooperate in investigating complaints of crop damage attributed to fumes from a plant operated in connection with the Manhattan Project."*
>
> *Signed L. R. Groves, Major General USA.*

"The Department of Justice is cooperating in the defense of these suits," wrote General Groves in a Feb. 28, 1946 memo to the Chairman of the US Senate Committee on Atomic Energy. Why the national security emergency over a few lawsuits by New Jersey farmers? In 1946 the United States had began full scale production of atomic bombs. No other nation had yet tested a nuclear weapon; and the A-bomb was seen as crucial for US Leadership of the post-war world. The New Jersey fluoride lawsuits were a serious roadblock to that strategy. "The specter of endless lawsuits haunted the military," writes Lansing Lamont in his acclaimed book about the first atomic bomb test "Day of Trinity."

In the case of fluoride "If the farmers won, it would open the door to further suits, which might impede the bomb program's ability to use fluoride." said Jacqueline Kittrell, a Tennessee public interest lawyer specializing in nuclear cases, who examined the declassified fluoride documents. (Kittrell has represnted plaintiffs in several human radiation experiment cases.) She added "The reports of human injury were especially threatening because of the potential for enourmous settlements—not to mention PR problems."

Indeed, du Pont was particularly concerned about the "possible psychologic reaction" to the New Jersey pollution incident, according to a secret 1946 Manhattan Project memo. Facing a thread from the Food and Drug Administration (FDA) to embargo the regions produce because of "high fluoride content" du Pont dispatched its lawyers to FDA offices in Washington, where an agitated meeting ensued. According to a memo sent next day to General Groves, du Pont's lawyer argued 'that in view of the pending suits . . . any action by the FDA . . . who would have a serious effect on the du Pont Company and would create a bad public relations situation." After the meeting adjourned, Manhattan Project Captain John Davies approached the FDA's Food Division chief and impressed upon Dr. White the substantial interest which the Government had in claims which might arise as a result of action which might be taken by the FDA. There was no embargo. Instead new tests for fluoride in the New Jersey area would be conducted, not by the Department of Agriculture—but by the US Army Chemical Warfare Services because "work by the Chemical Warfare Services would carry the greatest weights as evidence if . . . lawsuits are started by the complainants." the memo was signed by General Groves. Meanwhile, the public relations problem remained unresolved—local citizens in a panic about fluoride.

The farmer's spokesman, Willard B. Kille, was personally invited to dine with General Groves—then known as the man who built the atomic bomb—at his office of the War Department on March 26, 1946. Although he has been diagnosed with fluoride poisoning by the doctor, Kille departed the luncheon convinced of the government's good faith. The next day he wrote to the general, wishing the other farmers could have been present, he said, so "they too could come away with the feeling that their interest in this particular matter were being safeguarded by men of the highest type whose integrity they could not question." In a subsequent secret Manhattan Project memo, a broader solution to the public relations problem was suggested by chief fluoride toxicologist Harold C, Hodge. He wrote to the Medical Section Chief, Col. Warren, "Would there be any use in making attempts to counteract the local fear of fluoride on the part of the residents of Salem and Gloucester counties through lectures on Fluoride toxicology and perhaps the usefulness of Fluoride on health?" Such lectures were indeed given, not only to New Jersey citizens but to the rest of the nation throughout the Cold War.

The New Jersey farmer's lawsuits were ultimately stymied by the government's refusal to reveal the key piece of information that would have settled the case—how much fluoride du Pont had vented into the atmosphere during the war. "Disclosure would be injurious to the military security of the United States," wrote Manhattan Project Major C.A. Taney, Jr. The farmers were pacified with token financial settlements, according to interviews with descendants still living in the area. "All we knew is that du Pont released some chemical that burned up all the peach trees around here," recalls Angelo Giordano whose father James was one of the original plaintiffs. "The trees were no good after that, so we had to give up on the peaches." Their horses and cows, too, acted stiff and walked stiff, recalls his sister Mildred. "Could any of that have been the fluoride?" she asked. (The symptoms she detailed to the authors are cardinal signs of fluoride toxicity, accotding to veterinary toxicologists.) The Giordano family, too, has been plagued by bone and joint problems, Mildred adds.

Recalling the settlement received by the Giordanos, Angelo told these reporters that "my father said he got about $200." The farmers were stonewalled in their search for information, and their complaints have long since been forgotten. But they unknowingly left their imprint on history—their claims of injury to their health reverberated through the corridors of power in Washington, and triggered intensive secret bomb-program research on the health effects of fluoride. A secret 1945 memo from Manhattan Project Lt. Col. Rhodes to General Groves stated: "Because of complaints that animals and humans have been injured by hydrogen fluoride fumes in (the New Jersey) area, although there are no pending suits involving such claims, the University of Rochester is conducting experiments to determine the toxic effect of fluorides." Much of the proof of fluoride's safety in low doses rests on the post-war work performed by the University of Rochester, in anticipation of lawsuits against the bomb program for human injury."

Fluoride and the Cold War

"Delegating fluoride safety studies to the University of Rochester was not surprising. During World War II the federal government had become involved, for the first time, in large scale funding of scientific research at government owned labs and private colleges. Those early spending priorities were shaped by the nation's often secret-military needs. The prestigious upstate New York college, in particular, had housed a key wartime division of the Manhattan Project, studying the health of the new "special materials" such as uranium, plutonium, beryllium, and fluoride, being used to make the atomic bomb. That work continued after the war, with millions of dollars flowing from the Manhattan Project and its successor organization the Atomic Energy Commission (AEC). (Indeed, the bomb left an indelible imprint on all US science in the late 1940s and 1950s. Up to 90% of federal funds for university research came from either the Defense Department or the AEC in this period, according to Noam Chomsky's 1996 book "The Cold War and the University.")

The University of Rochester medical school became a revolving door to senior bomb program scientists. Postwar faculty included Stafford Warren, the top medical officer of the Manhattan Project, and Harold C. Hodge, chief of fluoride research for the bomb program. But this marriage of military secrecy and medical science bore deformed offspring. The University of Rochester classified fluoride studies-code-name Program F, were conducted at its Atomic Energy Project (AEP), a top secret facility funded by the AEC and housed in Strong Memorial Hospital. It was there that one of the most notorious human radiation experiments of the Cold War took place, in which unsuspecting hospital patients were injected with toxic doses of radioactive plutonium. Revelation of this experiment in a Pulitzer Prize—winning account by Eilean Wellsome led to a 1995 US Presidential investigation, and a multimillion dollar cash settlement for victims.

Program F was not about children's teeth. It grew directly out of litigation against the bomb program and its main purpose was to furnish scientific ammunition which the government and its nuclear contractors could use to defeat lawsuits for human injury. Program's F director was none other than Harold C. Hodge, who had led the Manhattan Project investigation of alleged human injury in the New Jersey fluoride pollution incident. Program's F purpose is spelled out in a classified 1948 report. It reads: "To supply evidence useful in the litigation arising from alleged loss of fruit crop several years ago, a number of problems have been opened. Since excessive blood fluoride levels were reported in human residents of the same area, our principal effort has been devoted to describing the relationship of blood fluorides to toxic effects."

The litigation referred to, of course, and the claims of human injury were against the bomb program and its contractors. Thus, the purpose of Program F was to obtain evidence useful in litigation against the bomb program. The research was being conducted by the defendants. The potential for conflict of interest is clear. If lower dose ranges were found hazardous by Program F, it might have opened the bomb program and its contractors to lawsuits for injury to human health, as well as public outcry. Comments lawyer Kittrell: "This and other documents indicate that the University of Rochester's fluoride research grew out of the New Jersey lawsuits and was performed to anticipation of lawsuits against the bomb program for human injury. Studies undertaken for litigation purposes by the defendants would not be considered scientifically acceptable today." adds Kittrell, "because of their inherent bias to prove the chemical safe."

Unfortunately, much of the proof of fluoride's safety rests on the work performed by Program F Scientists at the University of Rochester. During the postwar period that university emerged as the leading academic center for establishing the safety of fluoride, as well as its effectiveness in reducing tooth decay, according to Dental School spoke person William W. Bowen, M.D. The key figure in this research, Bowen said, was Harold C. Hodge—who also became a leading national proponent of fluoridating public drinking water. Program F's interest in water fluoridation was not just to counteract the local fear of fluoride on the part of residents, as Hodge had earlier written. The bomb program needed human studies, as they had needed human studies for plutonium, and adding fluoride to public water supplies provided one opportunity.

The A-bomb Program and Water Fluoridation Bomb program scientists played a prominent role in the nation's first planned water fluoridation experiment, in Newburgh, New York. The Newburgh Demonstration Project is considered the most extensive study of the health effects of fluoridation, supplying much of the evidence that low doses are safe for children's bones, and good for their teeth. Planning began in 1943 with the appointment of a special New York State Health Department committee to study the advisability of adding fluoride to Newburgh's drinking water. The chairman of the committee was Dr. Hodge, then chief of toxicity studies for the Manhattan Project. Subsequent members included Henry L. Barnett, a captain in the Project's Medical section, a John W. Fertig, who in 1944 was with the office of Scientific Research and Development, the Pentagon group, which sired the Manhattan Project. Their military affiliation were kept secret. Hodge was described as a pharmacologist, Barnett as a pediatrician. Placed in charge of the Newburgh project was David B. Ast, chief dental officer of the State Health

Department. Ast had participated in a key secret wartime conference on fluoride held by the Manhattan Project, and later worked with Dr. Hodge on the Project's investigation of human injury in the New Jersey incident, according to once-secret memos.

The committee recommended that Newburgh be fluoridated. It also selected the types of medical studies to be done, and "provided expert guidance" for the duration of the experiment. The key question to be answered was "Are there any cumulative effects—beneficial or otherwise, on tissues and organs other than teeth—of long continued ingestion of such small concentrations . . . ? According to the declassified documents, this was also key information sought by the bomb program, which would require long continued exposure of workers and communities to fluoride throughout the Cold War. In May 1945, Newburgh's water was fluoridated, and over the next ten years its residents were studied by the State Health Department. In tandem, Program F conducted its own secret studies, focusing on the amounts of fluoride Newburgh's residents retained in their blood and tissues—key information sought by the bomb program. "Possible toxic effect of fluoride were in the forefront of consideration," the advisory committee stated. Health Department personnel cooperated, shipping blood and placenta samples to the Program F team at the University of Rochester. The samples were collected by Dr. David B. Overton, the Department chief of pediatric studies at Newburgh.

The final report of the Newburgh Demonstration Project published in 1956 in the Journal of the American Dental Association concluded that "small concentrations" of fluoride were safe for US Citizens. The biological proof—based on work performed at the University of Rochester Atomic Energy Project—was delivered by Dr. Hodge.

Today, news that scientists from the bomb program secretly shaped and guided the Newburgh fluoridation experiment and studied the citizen's blood and tissue samples, is greeted with incredulity. "I am shocked—beyond words," said present day Newburgh Mayor Audrey Carey, commenting on these reporter's findings. "It reminds me of the Tuskegee experiment that was done on syphilis patients down in Alabama."

(This was another example of infecting some hospital patients with syphilis without their knowledge for experimental purposes.)

As a child in the early 1950s, Mayor Carey was taken to the old firehouse on Broadway in Newburgh, which housed the Public Health Clinic. There, doctors from Newburgh fluoridation project studied her teeth and a peculiar fusion of two finger bones on her left hand she had been born with. Today, adds Carey, her grandfather has white dental fluorosis marks on his front teeth. Mayor Carey wants answers from the government about secret history of fluoride, and the Newburgh fluoridation experiment. " I absolutely want to pursue it," she said. "It is appealing to do any kind of experimentation and study without people's knowledge and permission." Contacted by these reporters, the director of the Newburgh experiment, David B. Ast says he was unaware Manhattan Project scientists were involved. "If I had known, I would have been certainly investigating why, and what the connection was," he said. Did he know that blood and placenta samples from Newburgh were being sent to bomb program researchers at the University of Rochester? "I was not aware of it," Ast replied. Did he recall participating in the Manhattan Project's secret wartime conference on fluoride in January 1944, or going to New Jersey with Dr. Hodge to investigate human injury in the du Pont case—as secret memos state? He told the reporters he had no recollection of these events.

A spoke person for the University of Rochester Medical Center, Bob Loeb, confirmed that blood and tissue samples from Newburgh had been tested by the University's Dr. Hodge. On the ethics of secretly studying US citizens to obtain information useful in litigation against the A-bomb Program, he said, "that's a question we cannot answer." He referred inquires to the US Department of Energy (DOE). A spoke person for the DOE in Washington, Jayne Brady, confirmed that a review of DOE files indicated that a "significant reason for fluoride experiments conducted at the University of Rochester after the war was" impending litigation between the du Pont company and residents of New Jersey areas. "However," she added, "DOE has found no documents to indicate that fluoride research was done to protect the Manhattan Project involvement or its subcontractors from lawsuits." On Manhattan Project involvement in Newburgh, the spoke person stated, "Nothing that we have suggests that the DOE or predecessor agencies, especially Manhattan Project, authorized fluoride experiments to be performed in the 1940s. When told that the reporters had several documents that directly tied the Manhattan Project successor agency at the University of Rochester, the AEP (Atomic Energy Project), to the Newburgh experiment, the DOE spoke person later conceded her study was confined to "the available universe" of documents. Two days later, Jayne Brady faxed a statement for clarification, "My research only involved the documents that are collected as part of our radiation experiment project, fluoride was not part of our research effort." "Most significant" the statement continued, "the relevant documents may be in a classified collection at the DOE Oak Ridge National Laboratory, known as the Records Holding Task Group. This collection consists entirely of classified documents removed from other files for the purpose of classified document accountability many years ago," and was " a rich source of documents for the human radiation experiment project." she said. The crucial question arising from this investigation is, **Were adverse health findings from Newburgh and other bomb program fluoride studies suppressed?** All AEC funded studies had to be declassified before publication in civilian medical and dental journals. Where are the original classified versions?

The transcript of one of the major secret scientific conferences of WW2 on—"fluoride metabolism" . . . is missing from the files of the US National Archives. Participants in the conference included key figures who promoted the safety of fluoride and water fluoridation to the public after the war.

Harold C. Hodge, of the Manhattan Project, David B. Ast of the Newburgh Project, and US Public Health Service dentist H. Trendley Dean, popularly known as the "father of fluoridation." If it is missing from the files, it is probably still classified," National Archive librarian told these reporters.

A 1944 WW2 Manhattan Project classified report on water fluoridation is missing from the files of the University of Rochester Atomic Energy Project, the US National Archives, and the Nuclear

Repository at the University of Tennessee, Knoxville. The next four numerically consecutive documents are also missing, while the remainder of the "MP-1500 series" is present. "Either those documents are still classified, or they've been 'disappeared' by the government," says Clifford Honicher, Executive Director of the American Environmental Health Studies Project, in Knoxville, Tennessee, which provided key evidence in the public exposure and prosecution of US human radiation experiments. Seven pages have been cut out of a 1947 Rochester bomb project notebook entitled "Du Pont Litigation. Most unusual" commented chief medical school archivist Chris Hoolihan. Similarly, Freedom of Information Act (FOIA) requests by these authors over a year ago with the DOE for hundreds of classified fluoride reports have failed to dislodge any. "We're behind," explains Amy Rothrock, FIOA officer at the Department of Energy at their Oak Ridge operations.

Was information suppressed? These reporters made what appears to be the first discovery of the original classified version of a fluoride safety study by bomb-program scientists. A consorted version of this study was later published in the August 1948 Journal of American Dental Association. **Comparison of the secret with the published version indicates that the US AEC did censor damaging information on fluoride, to the point of tragicomedy. This was a study of the dental and physical health of workers in a factory producing fluoride for the A-bomb program, conducted by a team of dentists from the Manhattan Project. The secret version reports that most of the men had no teeth left. The published version reports only that the men had fewer cavities. The secret version says the men had to wear rubber boots because the fluoride fumes disintegrated the nails in their shoes. The published version omits this statement. The published version concludes that "the men were unusually healthy, judged from both a medical and dental point of view."**

Asked for comment on the early links of the Manhattan Project to water fluoridation, Dr. Harold Slavkin, Director of the National Institute for Dental Research, said, "I wasn't aware of any input from the Atomic Energy Commission." Nevertheless, he insisted, fluoride's efficacy and safety in the prevention of dental cavities over the last fifty years is well proved. The motivation of a scientist is often different from the outcome" he reflected "I do not hold a prejudice about where the knowledge comes from."

After comparing the secret and published versions of the consorted study, toxicologist Phyllis Mullenix commented, "This makes me ashamed to be a scientist." Of other Cold War era fluoride safety studies, she said, "Were they all done like this?"

Joel Griffiths is a medical writer in the New York City, author of a book on radiation hazards and numerous articles for medical and popular publications. Chris Bryson holds a master's degree from the Columbia University Graduate School of Journalism and has worked for the British Broadcasting Corporation and the Manchester Guardian.

I would like to overwhelmingly thank Joel Griffiths and Chris Bryson for some very intensive investigative work on the fluoride question. The conclusion that I came to is that powers that be continually are shoving this *silent killer* down our throats whether some of us want it or not. You, dear reader, can come to your own conclusion whatever that may be. There is in legal terms, a preponderance of evidence of the toxic effect fluoride has on numerous body functions.

http://www.healthy-communications.com/fluoridetoxicbomb.html

http://www.infiniteunknown.net/2010/10/31/fluoride-the . . .

Effects of Fluoride on Neurotransmitters in Fetal Brains

There were numerous studies done in China of the effect of fluoride on the fetal brains. The human placenta does not prevent the passage of fluoride from pregnant mother's bloodstream to the fetus. A fetus can be harmed by fluoride ingested by the mother during pregnancy. The fetal brain is one of the organs susceptible to fluoride poisoning. Four Chinese studies have investigated fluoride's effect on the fetal brain and each has found evidence of significant neurological damage, including neuronal degeneration and reduced levels of neurotransmitters, such as norepinephrine. Chinese researcher Yu (1996) stated that "when norepinephrine levels drop, the ability to maintain an appropriate state of activation in the central nervous system is weakened." Studies of fluoride treated animals have reported similar effects, including lower levels of norepinephrine.(Kaul 2009, Li 1994).

A. Yu Study

Two papers, provided by Yu and Dong, provide results of a single investigation of twenty fetuses—ten from high-fluoride area, endemic to the area, and ten from a low-fluoride area, in the water system. A third study was done from this investigative study, to determine fluoride effect on other bodily tissues. The report states, "The mothers of the ten fetuses that formed the subject group for this study, all had dental fluorosis, indicating that corresponding increase in urinary fluoride, indicating that these pregnant women were suffering from chronic fluoride poisoning. The excess fluoride of the mother was passed through the placental barrier into the fetus, and from there through the blood-brain barrier to accumulate in the fetal brain, leading to a significant rise in bone and brain fluoride levels. Our results are consistent with earlier reports. Previous experiments have shown that the brain of fetuses from endemic fluorosis areas as well as fluoride poisoned rats, manifest morphological changes. Following experimental testing of the monoamine neurotransmitters in fetuses from fluorosis endemic areas. The present study found lowered levels of norepinephrine and elevated levels of epinephrine. The presence of norepinephrine in the brain allows the organism to become alert, guards against the intensification of reflex reactions and other behaviors. Norepinephrine also plays a role in the regulation of complex response mechanisms, emotions, cerebro-cardio functions, etc. When norepinephrine levels drop, the ability to maintain an appropriate state of activation in the central nervous system is weakened. The elevated levels of epinephrine could be due to a blockage of the pathway that transforms epinephrine into norepinephrine or possibly due to suppression of the relevant metabolic enzymes, causing brain levels of epinephrine to increase, and the levels of norepinephrine to decrease" (Yu, Y., et al. 1996, "Neurotransmitter and receptor changes in the brains of fetuses from areas of endemic fluorosis," *Chinese Journal of Endemiology*, 15-257-259; one can access the entire study translated into English by going to the following title: "Neurotransmitter and Receptor Changes in the Brains of fetuses from Areas of Endemic Fluorosis").

B. Dong Study

It was determined that "the contents of five amino acid neurotransmitters and three types of monoamine neurotransmitters in the brains of fetuses aborted through induced labor in chronic fluorosis endemic areas. Findings revealed that the content of the excitatory amino acid, aspartic acid, was significantly lower, then in the fetuses from non-endemic area, whereas the content of the inhibitory amino acid, taurine, was significantly higher, the content of the major spinal cord-inhibitory glycine was significantly reduced. Among the monoamine neurotransmitters, the content of 5-hydroxytryptamine in the frontal and occipital lobes were elevated and the content of 5-hydroxytryptamine in the parietal lobe (pre-central and post-central gyri) was reduced" (Dong, Z., et al. 1993; "Determination of the contents of amino acid and mono-amino neurotransmitters in fetal brains from fluorosis endemic areas," *Journal of Guiyang Medical College* 1B(4); 241–245).

What this all means is that the amount of fluoride endemic to an area contributed to the imbalance of the necessary amino acids and mono-amino neurotransmitters in the brain during fetal development. Could this translate into potential cause of autism? The level of fluoride in the endemic fluorosis areas from urine analysis of these women was listed as: 4.37 ± 2.94 mg/liter. (One can access the entire study translated into English under following title: "Determination of the Contents of Amino Acid and Monoamine Neurotransmitters in Fetal Brains from Fluorosis Endemic Area.")

C. Du Study

"It is known that fluoride can cross the placenta from the mother's blood to the developing fetus. However, the theory here is a direct link between fluoride and brain cell damage is still controversial due to the lack of adequate evidence. In order to determine if there are any adverse effects on the developing human brain, especially starting from formation of the embryo, fetuses from an endemic fluorosis area at the 5th–8th month of gestation were compared to those from non-endemic area."

Results: "Normal Purkinje cells from the non-endemic fluorosis area were observed in single or parallel lines and were well organized in the fetal cerebellum. Purkinje cells of fetuses from the fluorosis endemic areas were abnormally disorganized and had a thicker, granulated layer in the cerebellum. Other dysmorphology, includes higher nucleus-cytoplasm ratio of brain cones, hippocampus cones and Purkinje cone cells, supports the theory that fluoride has an adverse effect on brain development. SEM analysis also found reduced neurons of brain cortex, decreased numerical density, volume density, and surface density in these fetuses from the endemic fluorosis area."

Summary: "The passage of fluorine through the placenta of mothers with chronic fluorosis and its accumulation within the brain of the fetus, impacts the developing central nervous system and stunts neuron development" (Du, L., 1992. "The effect of fluorine on the developing human brain," *Chinese Journal of Pathology* 21(4)218–220).

Need more proof? One can access the entire study translated into English under the following title: "The Effect of Fluoride on the Developing Human Brain."

D. Han Study

"In recent years, researches have noted that fluoride poisoning appears to begin in the fetal stage. Our study collected specimens from induced abortions in both fluoride endemic areas and non-affected areas, and by means of histochemical analysis, enzyme-chemical analyses, light microscopy, and electron microscopy, investigated the effects of fluoride on the fetus, providing evidence for early childhood contraction of fluorosis."

Results: "When the various hard and soft tissues taken from fetuses as part of this study were tested for fluoride, the results showed that the brain and bone tissue of the fluoride endemic area fetuses had higher fluoride content than the controls. The reason for this disparity is the previous excess fluoride intake of the mother. Fluoride can pass through the blood-brain barrier and accumulate in brain tissue. Thus in our study the brain tissue of the fetuses from the endemic area showed higher fluoride levels than the control. The mechanisms involved are not yet quite clear. Besides increased amounts of fluoride, the brain tissue of the endemic subjects also showed nerve cells with swollen mitochondria, expanded granular endoplasmic reticula, grouping of the chromatin, damage to the nuclear envelope, a lower number of synapses, fewer mitochondria, microtubules, and vesicles within the synapses; and damage to the synaptic membrane. These changes indicate that fluoride can retard the growth and division of the cells in the cerebral cortex. Fewer mitochondria, microtubules, and vesicles within the synapses could lead to fewer connections between neurons and abnormal synapses function, influencing the intellectual development after birth" (Han, H., et al. 1989. "Effect of fluorine on the human fetus," *Chinese Journal of Control of Endemic Diseases* 4:136–138; one can access the entire study translated into English under following title: "Effects of Fluorine on the Human Fetus.").

Does this imply autism?

E. Study of Fluoride's Effect on Liver, Thyroid Gland, and Adrenal Glands

The following study involves the same fetal tissue that was examined in the Yu (1996) and Dong (1993) studies. "This study examined fluoride's effect on the ultrastructure of cells in several tissues in the body, including from the thyroid gland. As can be seen in the following description, the fetuses from the fluoride exposed women were experiencing a systemic toxic effect . . . Objective, ultrastructural changes of epithelial cells of liver, adrenal glands, and thyroid glands of human fetuses from a fluorosis-endemic area were observed to provide the experimental basis for research into the mechanism of cellular damage caused by fluoride. Methods: 10 human fetuses in a fluorosis-endemic area were collected whose mothers all had dental fluorosis with urinary fluoride content of 4.37 plus or minus 2.94 mg/ liter. 10 human fetuses in a non-fluorosis-endemic area were collected whose mothers had no dental fluorosis with urinary fluoride content of 1.67 plus or minus 0.82 mg/liter. The fluoride electrode method was used to test the fluoride content in fetal bones. Tissues of livers, adrenal glands, and thyroid glands of the fetuses were taken for electron microscopic examinations."

Results: "Electron microscopic examinations showed: the major changes of cell membranes were microvilli that were shortened, reduced in number or even vanished. Fatter cellular connections were loose and their structure was disordered. Myelin-like structures were formed in those with severe pathological changes. The major mitochondrial changes were swollen mitochondria with increased volume, and even vanished and vacuolated cristae. The major pathological changes of endoplasmic reticulum were dilated and vesicular rough endoplasmic reticulum. The major pathological changes of cell nuclei were damaged and dilated, vesicular dual layer structure of nuclear membranes. Huge inclusion bodies or particles with relatively high abnormal electron density appeared in some cytoplasm."

Conclusion: "Fluorine damage to cell structure was multifaceted. Cell membrane, mitochondria, rough endoplasmic reticulum, and nuclear membranes could all be damaged at the time of fluorosis." Bottom line: creating abnormal cells have a negative effect on the proper activity of the liver, the thyroid gland, and the adrenal gland. Could this point to autism and hypothyroidism, affecting metabolic rate, resulting in obesity and related to improper function of the liver in removing toxins from the body? (Yu, Y. 2000. "Effects of fluoride on the ultrastructure of glandular epithelial cells of human fetuses," *Chinese Journal of Endemiology* 19(2): 81–83; one can access the entire study translated into English under following title: "Effects of fluoride on the ultrastructure of glandular epithelial cells of human fetuses").

F. Harvard Study on Fluoridated Water's Effect on IQ

A recent Harvard University study funded by National Institute of Health (NIH) shows that children that live in areas in highly fluoridated water have significant lower IQ scores than those who live in low fluoride areas. NRC (National Research Council) reports that high concentrations of fluoride in drinking water may cause neurotoxicity in lab animals, including effect on learning and memory. Findings from meta-analyses of twenty-seven studies suggest an inverse association between high-fluoride exposure and children's intelligence. The results suggest that fluoride may be a developmental neurotoxicant that affects brain development at exposures much below those that can cause toxicity in adults. (This is a fact determined by the Yu, Dong, Du, and Han studies. It is not a maybe but a *fact*. Damage occurs in fetal stages due to fluoride present in the baby formulas while their brain is still developing. Can autism be far behind?) Rats exposed to 1 ppm of water soluble fluoride for one year showed morphological alterations in the brain and increased levels of aluminum in brain tissue. (Is this a potential cause of increased Alzheimer's disease in adults?)

There are so many scientific studies showing the direct toxic effects of fluoride on our bodies. Despite the evidence against it, fluoride is still added to 70% of US drinking water. It is amazing that the medical and dental communities are so stubbornly resistant to connect the dots when it comes to skyrocketing increase of cognitive decline in adults and behavioral issues in children like autism, ADD, ADHD, depression, and learning disabilities of all kinds There have been over twenty-three human studies and one hundred animal studies linking fluoride toxicity to brain damage.

Reported effects on fluoride on brain include reduction in nicotinic acetylcholine receptors, damage to one's hippocampus, formation of beta-amyloid plaques (the classic brain abnormality in Alzheimer's disease), reduction in lipid content resulting in damage to Purkinje cells, exacerbation of lesions induced by iodine deficiency, impaired antioxidant defense system resulting in increased uptake of aluminum, and also accumulation of fluoride in the pineal gland.

http://www.chinesefluorideeffectstudiesonhumanfetalbrains.com

Neurotoxicity and Neurobehavioral Effects of Fluoride

Fluoride in Drinking Water: A Scientific Review of EPA's Standards

This chapter of NRC report reviews the effect of fluoride on the central nervous system and behavior. The human data includes epidemiological studies of populations exposed to different fluoride concentrations. Laboratory studies of behavioral, biochemical, and neuroanatomical changes induced by fluoride were reviewed and summarized. Let's first take a look at the cognitive studies by a number of Chinese scientists.

Cognitive studies

Several Chinese studies have reported the effects of fluoride in drinking water on cognitive capacities (X., Li et al.1995; Zhao, et al. 1996; Lu et al. 2000; Xiang et al. 2003a,b). The strongest design study by Xiang 2003a, compared the intelligence of 512 children (ages 8–13) living in two villages with different fluoride concentrations in the water. The high fluoride concentration in the village of Wamaio was 2.47 ± 0.79 mg/liter, and the low fluoride concentration in the village of Xinhuai was 0.35 ± 0.15 mg/liter. The population had comparable iodine and creatine concentrations, family income, and educational levels, as well as other social factors. The population was only exposed to fluoride water but not to smoke from coal fires, industrial pollution, or consumption of brick tea. (These are other sources of fluoride.) The mean urine fluoride concentration found in the Wamaio children was 3.47 ± 1.95 mg/liter, and in children from Xinhuai village 1.11 ± 0.39 mg/liter. Using the combined Raven's test for rural China, the average intelligent quotient (IQ) of the children in Wamaio was found to be significantly lower (92.2 ± 3.0), ranging 54–126, than that of Xinhuai children (100.41 ± 13.21), ranging 60–128. The IQ scores in both male and female children declined with increasing fluoride exposure. Modal scores of the IQ distribution in the two villages were approximately the same.

A comparable study was done by Lu et al. 2000, in a different area of China, also compared IQs of 118 children, ages 10–12 years, living in two different areas with different fluoride concentrations in the water (3.15 ± 0.61 mg/liter versus 0.37 ± 0.04 mg/liter). Children had similar social and educational levels. The combined Raven's tests showed significant lower mean IQ scores of children in the high-fluoride area (92.27 ± 20.45 mg/liter) than the children in low-fluoride area (103 ± 13.86 mg/liter). Of special importance is that 21.6% of the children in high-fluoride village scored 70 or below on the IQ scale, while children in the low-fluoride village, only 3.4% had low scores.

Zhao et al. 1996 also compared the IQs of 160 children, ages 7–14, living in high- and low-fluoride villages. Using the Chinese Rui Won test, the investigation found that the average IQ of children in high-fluoride area on average was 97.69, significantly lower than that of children in the low-fluoride area, on average was 105.21. The investigations also reported that enamel fluorosis was present in 86% of the children in the high-fluoride concentrations in the water.

Another Chinese study evaluated fluoride exposure to inhalation of soot and smoke from domestic coal fires used for cooking and drying grain (Li et al. 1995). Many of the children exhibited moderate to severe dental fluorosis. The average IQ of 900 children from severe dental fluorosis was 9–11 points lower than the children with low or no dental fluorosis.

The NRC in their analysis tries to discredit the Chinese studies as to test procedural details. They try to explain that many factors outside of intelligence influence performance on IQ tests. Well, how do you explain the significant difference in the IQ levels between the two different fluoride levels that were tested? I beg to differ with the NRC explanation.

Following is a copy of the news release regarding the Harvard study:

Harvard Study Finds Fluoride Lowers IQ

(Published in a federal government journal, New York, July 24, 2012/PRNewswire-USNewswire): Harvard University researchers' review of fluoride/brain studies concludes "our results support the possibility of adverse effects of fluoride exposure on children's neurodevelopment." It was published online, July 20, in Environmental Health Perspectives. A US National Institute of Environmental Health Science journal reports the NYS Coalition Opposed to Fluoridation Inc. (NYSCOP): "The children in high fluoride areas had significant lower IQ than those who lived in low fluoride area," writes Choi et al.

The problem I have with this release is the wording "possibility of neurodevelopment." The Chinese studies were confirmed by the Harvard study as to the lower IQ results. The other Chinese studies found cellular damage to the brain. This was not a possibility but a *fact*. There are neurodevelopment effects on the brain as early as in the womb of the mother.

Mental and physiological changes

There are numerous reports of mental and physiological changes after exposure to fluoride from various sources—air, food, and water. There are a number of experimental studies of individuals who underwent withdrawal from their fluoride exposure and subsequent re-exposure under so-called blind conditions. In most cases, symptoms disappeared with elimination of fluoride exposure and returned when re-exposed to fluoride. These observations of these individuals suggested that fluoride could be associated with cerebral impairment. Fluoride exposed people of all ages showed common symptoms of lethargy, weakness, and impaired ability to concentrate (this symptom is common to ADD, ADHD, and autism), regardless of the source of fluoride exposure. In some cases (50%) memory problems were reported (Spittle 1994), and also were described biochemical changes in enzymatic systems that could influence the physiological changes found in patients. The high exposure to fluoride, due to its high chemical reactivity, disrupts the N-H (nitrogen-hydrogen) bonds in amines of proteins, by the production of N-F (nitrogen-fluoride) bond (Emsley et al. 1981). This unnatural bond distorts the structure of a number of proteins with the potential to cause important biological effects. "Fluorides also distort the structure of cytochrome-C peroxidase" (Edwards et al. 1984). (Cytochrome-C oxidase is an enzyme affected by fluoride.) Spittle also noted the likelihood of fluoride interfering with the basic cellular energy sources used by the brain through the formation of aluminum fluoride (Jope et al. 1998) and subsequent effects on G-proteins (enzymes that act as molecular switches, transmitting signals). Aluminum and fluoride have the highest affinity for binding to each other.

Effect of silicofluoride

One of the fluoride compounds added to drinking water supply is silicofluoride. Silicofluoride (Westerndorf et al. 1975) were found to have greater influence in inhibiting the synthesis of cholinesterases, including acetylcholinesterase, than sodium fluoride. This produces a situation in which acerylcholine (ACh) accumulates in the vicinity of ACh terminals and leads to excessive activation of cholinergic receptors in the central and peripheral nervous systems (Knappwost and Westendorf 1974)—in other words, disruption of proper brain operation, such as electrical signals between neurons. Silicofluorides at high concentrations are used in insecticides and nerve gases due to this capability.

Dementia

For a number of decades (thirty), it has been known that Alzheimer's disease is associated with a substantial decline in cerebral metabolism (Sokoloff 1966). The decrease is reflected in the brain's metabolic rate of glucose, cerebral rate of oxygen, and cancelled blood flow, the reduction found in Alzheimer's disease patients, which is about three times greater than in patients with dementia. The temporal, parietal, and frontal regions of the brain are areas with some of the greatest reduction of aerobic metabolism (Weiner et al. 1993). One system specifically sensitive to carbohydrate utilization is the collection of areas involved with the synthesis of acetylcholine. The release of this neurotransmitter is negatively affected by the interruption of aerobic metabolism (Johnson et al. 1988; Silverman and Small 2002).

"Most of the drugs used today to treat Alzheimer's disease are agents that enhance the effects of the remaining acetylcholine system. Certain characteristics of Alzheimer's disease is a general reduction of aerobic metabolism in the brain. This translates in a reduction in energy available for neuronal and muscular activity."

A study by Jocqmin et al. 1994 in France says, **"a significant decrease in cognitive abilities was found when their drinking water contained calcium**, **aluminum, and fluorine."** According to the NRC report, **"the possibility exists that chronic exposure to aluminum fluoride can produce aluminum inclusions with blood vessels. The aluminum deposits inside the vessels and those attached to the intima could cause turbulence in the blood flow and reduced transfer of glucose and oxygen to the intercellular fluids."** (Intima is the innermost layer of an artery or vein. It is made up of one layer of endothelial cells.)

Animal studies

Dr. Phyllis Mullinex study on animals, rats, reported a link between fluoride and behavior. This was the first attempt by a US scientist, other than Dr. Horace C. Hodge during the Manhattan Project, to do a study of fluoride effect on the central nervous system. The study provided evidence that exposure to fluoride in rats, in prenatal, weaning, and adulthood, has affected the rats' behavior. Her findings were contrary to the beliefs held to that point. Further requests for funds for additional studies she wanted to do were met with a total runaround and denial by numerous federal agencies. Her study was criticized as to the method she used instead of focusing on the results.

Anatomy

Another study by Varner et al. (1994) reported that "all groups subjected to aluminum fluoride had significant losses of cells in the hippocampus area. The losses were not dose dependent. The two types of cellular anomalies were found in the treated animals: 1) argentophilic cells throughout the hippocampus and dentate gyrus with considerable sparing of cells in the CA2 region, and 2) increased aluminum in the inner and outer linings of a large number of blood vessels, cells containing aluminum inclusions were not uncommon. This enhancement of aluminum deposits is not surprising because the amount of aluminum found in the brain was almost double that found in control animals."

Another study undertaken by Varner et al. (1998) determined that the brains of both aluminum fluoride and sodium fluoride were twice the amount of control group, as well as aluminum content in kidneys and liver. This finding was supported by a study made by Strunecka et al. (2002) that fluoride enhances aluminum uptake, but it is organ specific.

Neurochemical effects and mechanisms

Biochemical changes in the brain were also studied. Guan et al. (1998) study reported alterations in the phospholipid content of rats exposed to sodium fluoride and aluminum fluoride. The consequence to these chemical exposures is an increase of free radicals. G-protein receptors mediate the release of many neural transmitters. They also are involved in regulating important systemic influences on the brain and behavior. Aluminum fluoride is involved in regulating the pineal hormone melatonin system as well as the thyroid hormone—growth hormone connection. "It is said that in this regard every molecule of aluminum fluoride is the messenger of false information" (Strunecka and et al. 2002, p. 275). Acetylcholine is especially important neurochemical transmitter that reaches all areas of the brain. Sodium fluoride and aluminum fluoride change the number of acetylcholine receptors in the brain of the rat (Long et al. 2002). Decreased number of nicotinic acetylcholine receptors, subunits, and nicotinic receptors have been related to Alzheimer's disease. The number of receptors for acetylcholine has been found reduced in regions of the brain thought to be most important for mental ability and for adequate memory ability. "G-protein mediate the release of many of the best known transmitters of the central nervous system. Fluorides affect transmitter concentrations and functions, but also are involved in the regulation of glucagons (a peptide hormone produced by alpha cells of the pancreas that raises the concentration of glucose in bloodstream), prostaglandins (are produced by the body, are hormone like substances in response to tissue damage or infection), and a number of central nervous system peptides, including vasopressin (a neurohypophysical hormone to control water retention), endogenous opioids (a narcotic type neurotransmitter), and other hypothalamic peptides. Fluoride increases the production of free radicals in the brain through several different biological pathways."

Per NRC's statement, **"On the basis of information largely derived from histological, chemical, and molecular studies, it is apparent that fluorides have the ability to interfere with the function of the brain and the body by direct or indirect means."**

Instead of strongly recommending cessation of water fluoridation and adding fluoride to mouthwashes and cessation of fluoride gel treatments by the dentists, NRC is recommending further studies. Is there an ulterior motive to continue fluoridation? (Nazi Germany used fluoridation to dumb down the population.)

Chapter 3

Endocrinology

The endocrine system is a collection of glands, each of which secretes different types of hormones that regulate metabolism, growth and development, tissue function, sexual function, reproduction, sleep, and mood, among other functions. The endocrine system is made up of eight major glands, which are groups of cells that produce and secrete chemicals in the form of hormones. A gland selects and removes materials from the blood, processes them, and secretes the finished product for use somewhere else in the body. Almost every organ and cell in the body is affected by the endocrine system. A group of glands that signal each other in sequence are usually referred to as an axis. One example is the hypothalamic-pituitary-adrenal gland axis, which coordinates interactions among hypothalamus, the pituitary gland, and the adrenal gland, also called suprarenal glands, which are small conical organs on top of the kidneys. The endocrine system sends signals throughout the body, much like the nervous system, but unlike the immediate responses triggered by the nervous system, the effects can take a few hours or weeks. Hormones released from the endocrine glands into the bloodstream travel to their target tissue to elicit a response. These hormones leave the gland and are transported to organs and tissues in every part of the body.

Endocrine glands are vascular and generally do not have ducts, using intracellular vacuoles, or granules, to store hormones. They differ from exocrine glands—salivary glands, sweat glands, and glands within the gastrointestinal tract—which have ducts or hollow lumen. The endocrine system gets some help from organs such as kidney, liver, heart, and gonads, which have secondary endocrine functions. The kidney, for example, secretes hormones, such as erythropoietin and renin. Hormone levels that are too high or too low are an indication of a problem with the endocrine system. Hormone diseases also occur if your body does not respond to hormones in the appropriate ways. Stress, infection, and changes in the blood's fluid and electrolyte balance can also influence hormone levels. The most common endocrine disease in the United States is diabetes, a condition in which the body does not properly process glucose. This is due to lack of insulin, or if the body is producing insulin, it is not working effectively. Hormone imbalances can have a significant impact on the reproductive systems, particularly on women. Endocrinologists treat patients with fertility issues and also assess and treat patients with health concerns surrounding menstruation and menopause.

Endocrinology is a specialty of internal medicine, which deals with the diagnosis and treatment of diseases related to hormones. Endocrinology covers such human functions as the coordination of metabolism, respiration, sensory perception, and movement. Endocrinology also focuses on the endocrine glands and tissues that secrete hormones. Hypothyroidism occurs when the thyroid gland does not produce enough hormones to meet the body's need. Insufficient thyroid hormone can cause many of the body's functions to slow or shut down completely. Thyroid cancer begins in the thyroid gland and starts when cells in the thyroid begin to change, grow uncontrollably, and eventually form a tumor. Hypoglycemia, also called low blood glucose or low blood sugar, occurs when blood glucose drops below normal levels. This typically happens as a result of treatment for diabetes when too much insulin is taken. It can occur in people not undergoing treatment for diabetes, but it is fairly rare. A metabolic disorder occurs when there is an imbalance of substances needed to keep the body functioning—hormone levels may be too high or low. Metabolic disorders happen when some organs, such as one's liver or pancreas, become diseased or do not function normally. Diabetes is an example. The bones can be impacted by hormones.

Osteoporosis and osteomalacia (rickets) can cause bones to soften and come under the guise of endocrine problems. The endocrine system consists of a number of glands. These glands produce and secrete hormones which control the body's metabolism, growth, sexual development, and function.

The main endocrine glands are as follows:

1) Hypothalamus, pituitary, thymus, and pineal glands, located in the head
2) Thyroid and parathyroid glands, located in the neck and upper chest
3) Pancreas and adrenals, located on top of kidneys and upper abdomen
4) Ovaries in females and testes in males

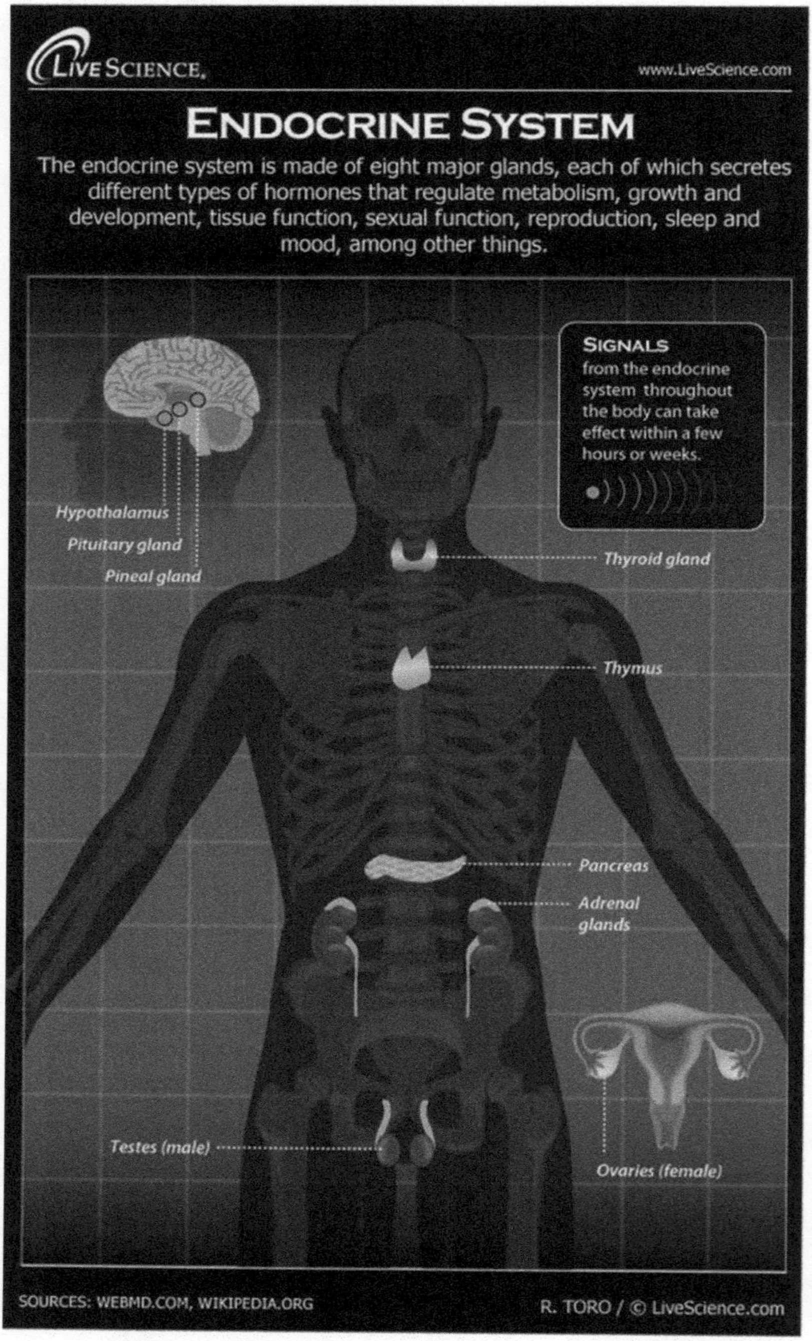
LIVE SCIENCE.
www.LiveScience.com
ENDOCRINE SYSTEM
The endocrine system is made of eight major glands, each of which secretes different types of hormones that regulate metabolism, growth and development, tissue function, sexual function, reproduction, sleep and mood, among other things.
SIGNALS
from the endocrine system throughout the body can take effect within a few hours or weeks.
Hypothalamus
Pituitary gland
Pineal gland
Thyroid gland
Thymus
Pancreas
Adrenal glands
Testes (male)
Ovaries (female)
SOURCES: WEBMD.COM, WIKIPEDIA.ORG
R. TORO / © LiveScience.com

The hypothalamus gland

Hypothalamus gland is located just above the brain stem, below the thalamus. This gland activates and controls involuntary body functions, appetite, sleep, temperature, as well as the sleep-wake cycle known as the circadian cycle. The cycle pertains to certain biological activities that occur at a particular interval regardless of constant rhythmic biological cycles recurring at approximately twenty-four-hour interval. The hypothalamus links the nervous system to the endocrine system via the pituitary gland. The hypothalamus is in control of pituitary hormones by releasing the following types of hormones: (a) thyrotropic-releasing hormone, stimulating the pituitary gland to release TSH (thyroid-stimulating hormone); (b) growth-releasing hormone, triggering the release of growth hormone from the pituitary gland; (c) corticotropin-releasing hormone, triggering the release of ACTH (adrenocorticotropic hormone) to stimulate the adrenal gland to secrete its hormones; (d) gonadotropin-releasing hormone, triggering the release of LH (lutein hormone) and FSH (follicle-stimulating hormone).

The thymus gland

Thymus gland secretes hormones commonly referred to as humoral factors and are important during puberty. The main role is to ensure a person develops a healthy immune system. It produces T-lymphocytes, which mature within the gland as part of the immune system.

The pineal gland

The pineal gland releases melatonin which helps the body to recognize when to go to sleep. It regulates the sleep-wake cycle.

The pituitary gland

The pituitary gland is located just off the hypothalamus at the base of the brain. It is also known as the main endocrine master gland. It secretes hormones that regulate the functions of other glands as well as growth and other several body functions. If it is underactive, hyposecretion of hormones, it can lead to dwarfism, slow bone growth in childhood, and underactive in other endocrine glands. If it is overactive, hypersecretion of hormones, it may lead to gigantism. The front part, anterior pituitary, produces the following hormones: (a) growth hormone, which promotes growth in children; for adults, maintains healthy muscle and bone mass; (b) prolactin, which stimulates milk production; low levels in males are linked to sexual problems; (c) adrenocorticotropic hormone (ACTH), which promotes the production of cortisol, which helps to reduce stress and maintain healthy blood pressure; (d) thyroid-stimulating hormone (TSH), which helps to regulate thyroid gland function, important in maintaining healthy metabolism throughout the body; (e) luteinizing hormone, which regulates estrogen in women and testosterone in men; (f) follicle-stimulating hormone (FSH) found in both males and females, which in women regulates releasing of eggs, and in men ensures normal function of sperm production.

The pituitary gland—back part or posterior pituitary secretes two hormones: (a) oxytocin, which causes pregnant women to start having contraction at the appropriate time and promote milk flow in nursing women; (b) antidiuretic hormone referred to as vasopressin, which helps to regulate water balance in the body.

The thyroid gland

Thyroid gland is located just below Adam's apple in the neck. It produces hormones that are highly significant. The thyroid gland consists of two types of cells called follicular and parafollicular cells. The follicular cells secrete two hormones, thyroxine (T4) and triiodothyronine (T3). Of the two hormones, the T3 is the more active one. When necessary, the T4 is converted to T3 by the liver.

The parafollicular cells secrete the hormone calcitonin. The thyroid gland needs iodine and tyrosine, an amino acid. The thyroid gland cells are the only cells in the body that can absorb iodine. The T3 and T4 hormones enter the body cells and control metabolism, where oxygen and nutrients are converted into energy. The hormone calcitonin regulates calcium metabolism in the body. It controls levels of calcium and phosphorus and prevents the body from breaking down the calcium in the bones.

The adrenal gland

The adrenal gland, located atop the kidneys, is divided into two regions. The right gland is triangular, while the left one is semilunar in shape. These glands secrete the following hormones, corticosteroids and catecholamine, such as norepinephrine and adrenaline (epinephrine), which are released during stress. Glandular hypersecretion, too much, may lead to overnervousness, raised blood pressure, and Cushing's disease (prolonged exposure to cortisol). Adrenal gland hyposecretion, too little, may lead to Addison's disease (insufficient adrenocortical insulin), mineralocorticoid deficiency, and diabetes. The adrenal glands also produce androgens, male sex hormones that promote the development of male characteristics. Testosterone is the major androgen. These glands also produce aldosterone, which affects kidney functions.

The ovaries and testicles

The testicles, located in the scrotum below the penis in males, secrete androgens, mainly testosterone, which control sexual development, puberty, facial hair, sexual behavior, libido, erectile function, and the formation of spermatozoa. The ovaries secrete two hormones, estrogen and progesterone. Estrogen is essential in a female's puberty, promoting the development of breasts, uterus, and other female characteristics. Progesterone is essential during ovulation and pregnancy. It also regulates the monthly menstrual cycle. Ovaries' hypersecretion, too much, leads to exaggerated female traits. Testicle hypersecretion, to much, lead to exaggerated male characteristics.

The pancreas

Pancreas, located in the abdomen, is both an endocrine gland and a digestive organ. It produces insulin, somatostatin, glucagon, and pancreatic polypeptides. Insulin plays a major role in carbohydrate and fat metabolism in the body. Somatostatin regulates endocrine and nervous systems function. It inhibits the secretion of several hormones, such as gastrin, insulin, and growth hormone. Glucagon is a peptide hormone which raises blood glucose levels when they fall too low. Pancreatic polypeptides help to control the secretion of substances made by the pancreas. A peptide is a molecule that is made up of at least two amino acids. Pancreatic hypersecretion may lead to hyperinsulinism, too much insulin, and not enough glucose to the brain. Hyposecretion, too little, leads to diabetes.

The parathyroid glands

Parathyroid glands are small endocrine glands located in the neck. They produce parathyroid hormone, which regulates calcium and phosphorus in the blood, blood clotting, and neuromuscular excitation.

As one can deduce from the above minimal descriptions of the endocrine system, there is a very delicate balance in the body. All the glands act synergistically to maintain proper life functions. Any very slight imbalance in the endocrine system will result in mental or physical malfunction in some form. There are built-in checks and balances to inhibit overproduction of certain hormones or increase hormone production if necessary. The problem arises when external influence overloads the endocrine system due to toxins or other immune disrupting elements, such as bacteria, viruses, or fungi, and also chemical toxins. With this in mind, let's examine what some of these external disruptors are on some of the endocrine glandular system, especially toxins.

http://www.medicalnewstoday.com/articles/248679
http://www.en.wikipedia.org/wiki/endocrinology

Fluoride Effect on the Pituitary Gland

The pituitary gland is called the master gland of the endocrine system because it controls the function of the other endocrine glands. The pituitary gland is the size of a pea and is located at the base of the brain. The gland is attached to the hypothalamus, part of the brain that affects the pituitary gland by nerve fibers. The gland consists of three distinct lobes: (1) anterior lobe, (2) the intermediate lobe, and (3) posterior lobe. Each lobe of the pituitary gland produces certain hormones. The anterior lobe produces the following hormones: growth hormone; prolactin to stimulate milk production after giving birth; adrenocorticotropin (ACTH) to stimulate the adrenal gland; thyroid-stimulating hormone (TSH) to stimulate the thyroid gland; follicle-stimulating hormone (FSH) to stimulate the ovaries and testes; luteinizing hormone (LH) to stimulate the ovaries and testes. The last two hormones are very essential in reproduction. The intermediate lobe secretes melanocyte-stimulating hormone to control skin pigmentation. The posterior lobe secretes antidiuretic hormone (ADH) to increase absorption of water into the blood by the kidneys; oxytocin to contract the uterus during childbirth and stimulate milk production.

http://www.umm.edu/endocrine/pitgland.htm)

http://www.fluoridealert.org/.../pesticides/effects.endocrine.pituitary.htm

Fluoride Effect on the Pineal Gland

In 1990s, British scientist, Jennifer Luke, discovered that fluoride accumulates in high levels in the pineal gland (Luke 2001). Dr. Jennifer Luke's discovery:

1) The pineal gland absorbs more fluoride than any other part of the body.
2) Total absorption can be up to 21,000 ppm (parts per million).

3) At this level, the hard tissue of the pineal gland absorbs a much larger amount of fluoride than any other hard tissue of the body.
4) The fluoride accumulation in the pineal gland restricts the production of melatonin and accelerates the onset of puberty.

The pineal gland is responsible for the synthesis and secretion of the hormone melatonin. Melatonin maintains the body's sleep-wake cycle, regulates the onset of puberty in females, and helps to protect the body from cell damage from free radicals. Preliminary animal experiments found that fluoride reduced melatonin levels had shortened the time to puberty. NRC (National Research Council) has stated that **"fluoride is likely to cause decreased melatonin production, and to have other effects on normal pineal function, which in turn could contribute to a variety of effects in humans"** (NRC 2006, p.256). When the production of melatonin is suppressed, then one is at risk for a number of conditions such as Alzheimer's disease, dementia, bipolar disease, insomnia, lower back pain, hormone imbalance, and pineal gland calcification. The fluoride content can come from many sources: fluoridated drinking water, store-bought products made with fluoridated water, medicine like Prozac and fluoroquinolone antibiotics, and nonstick cookware containing Teflon (Dr. Jennifer Luke, at University of Surrey in England, 1997).

In the US, the children are reaching the age of puberty at earlier ages than in the past. This raises a heightened risk for breast cancer. Results of the Newburgh, NY, fluoridation of their water supply system study, as early as the late and early 1940s–1950s, found that the girls living in fluoridated community reached puberty five months earlier than girls living in nonfluoridated community (Schlesinger 1956).

Melatonin levels control estrogen, progesterone, and testosterone amounts, which in turn signal menses (the monthly flow of blood and cellular debris from the uterus in women) and affect the start of menopause. Melatonin also affects the aging and immune systems, due to the sleep patterns, and the antioxidant effect of this hormone.

References:

Luke J. (2001). Fluoride deposition in the aged human pineal gland. Carries Res. 35(2)125–128.

National Research Council (2006). Fluoride in Drinking Water: A Scientific Review of EPA's Standards. Nat. Acad. Press Washington, DC.

Schlesinger, E. R., et al. (1956). Newburgh-Kingston caries fluorine study XIII Pediatric findings after ten years J. Amer. Dental Assoc. 52(3):296–306.

Estrogen signaling pathways. A link between breast cancer and melatonin actions. Cancer Detection and Prevention. Vol. 30, (2) p.118–128, 2006 by S. Cos, A. Gonzales, C. Martinez-Campa, M. Mediavilla, C. Alonso-Gonzales, E. Sanchess-Barcelli

http://www.fluoridealert.org/issues/health/pine

http://www.articles.mercola.com/.../fluoride-and-pineal-gland

http://www.dwcalcifypinealgland.com/dangers-of-fluoride

http://www.thearrowsoftruth.com/the-pineal-gland-fluoride-2012

Fluoride Effect on the Thyroid Gland

There are a great number of studies done on the destructive effect of fluoride on the thyroid gland that affects our metabolic system. One of the better articles regarding the effects of fluoride on the thyroid gland was written by Dr. Barry-Durrant-Peatfield MBBS, LRCP, MRCS, medical advisor to thyroid, UK (9-9-4). He best summarizes the toxic effect of fluoride on the thyroid gland.

There is a daunting amount of research studies showing that the widely acclaimed benefits on fluoride dental health are more imagined than real. My main concern however, is the effect of sustained fluoride intake on general health. Again, there is a huge body of research literature on this subject, freely available and in the public domain.

But this body of work was not considered by the York Review when their remit was changed from "Studies of the Effects of fluoride on health' to 'Studies on the effects of fluoridated water on health'. It is clearly evident that it was not considered by the BMA (British Medical Association), BDA (British Dental Association), BFS (British Fluoridation Society), and FPHM (Faculty for Public Health and Medicine), since they all insist, as in the briefing paper to Members of Parliament—that fluoridation is safe and non-injurious to health.

This is a public disgrace. I will now show by reviewing the damaging effects of fluoridation with special reference to thyroid illness. It has been known since the latter part of the 19th century, that certain communities, notably Argentina, India and Turkey were chronically ill, with premature aging, arthritis, mental retardation, and infertility; and high levels of natural fluorides in the water were responsible. Not only was it clear that the fluoride was having a general effect on the health of the community, but in the early 1920's Goldenberg, warning with Argentina showed that fluoride was displacing iodine; thus compounding the damage and rendering the community also hypothyroid from iodine deficiency.

Highly damaging to the thyroid gland. This was the basis of the research jn the 1930s of May, Litzka, Gorlitzer von Mundy, who used fluoride preparations to treat overactive thyroid illness. Their patients either drank fluoridated water, swallowed fluoride pills or were bathed in fluoridated bath water; and their thyroid function was, as a result, greatly depressed. The use in 1937 of fluorotyrosine for this purpose showed how effective this treatment was; but the effectiveness was difficult to predict and many patients suffered total thyroid loss. So it was given a new role and received a new name, Pardinon. It was marketed not for overactive thyroid disease, but as a pesticide. (Note the manufacturer of fluorotyrosine was IGFarben, who also made sarin, a gas used in World War II).

This bit of history illustrates the fact that fluorides are dangerous in general and in particular highly damaging to the thyroid gland, a matter to which I shall return shortly. While it is unlikely that it will be disputed that fluorides are toxic—let us be reminded that they are Schedule 2 POISONS under the Poisons Act 1972, the matter in dispute is the level of toxicity attributable to given amounts; in today's context the degree of damage caused by given concentrations in the water supply. While admitting its toxicity, proponents rely on the fact that it is diluted and therefore, it is claimed, unlikely to have deleterious effects. They could not be more mistaken.

It seems to me that we must be aware of how fluoride does its damage. It is an enzyme poison. Enzymes are complex protein compounds that vastly speed up biological chemical reactions while themselves remaining unchanged. As we speak, there occurs in all of us a vast multitude if these reactions to maintain life and produce the energy to sustain it. The chains of amino acids that make up these complex proteins are linked by simple compounds called amides; and it is with these that fluorine molecules react, splitting and distorting them, thus damaging the enzymes and their activity. Let it be said at once, this effect can occur at extraordinary low concentrations; even lower than the one part per million which is the dilution proposed for fluoridation in our water supply. The body can only eliminate half. Moreover, fluorides are accumulative and buildup steadily with ingestion of fluoride from all sources, which include not just water but the air we breathe and the food we eat. The use of fluoride toothpaste in dental hygiene and the coating of teeth are further sources of substantial levels of fluoride intake. The body can only eliminate half of the total intake, which means that the older you are, the more fluoride will have accumulated in your body. Inevitably, this means the ageing population is particularly targeted. And even worse for the very young there is a major element of risk in baby formula made with fluoridated water. The extreme sensitivity of the very young to fluoride toxicity makes this unacceptable. Since there are so many sources of fluoride in our everyday living, it will prove impossible to maintain an average level of 1 ppm as is suggested.

What is the result of these toxic effects?

"First, the immune system. The distortion of protein structure causes the immune protein to fail to recognize body proteins, and so instigates an attack on them, which is Autoimmune Disease. Autoimmune Disease constitute a body of disease processes troubling many thousands of people; Rheumatoid Arthritis, Systemic Lupus Erythematosus, Asthma and Systemic Sclerosis are examples, but in my particular context today, thyroid antibodies will be produced which will cause Thyroiditis resulting in the common hypothyroid disease, Hashimoto's Disease and the hyperthyroidism of Grave's Disease. Muscular Skeletal damage results further from the enzyme toxic effect; the collagen tissue of which muscles, tendons, ligaments and bones are made, is damaged. Rheumatoid illness, osteoporosis and deformation of bones inevitably follows. This toxic effect extends to the ameloblasts making tooth enamel, which is consequently weakened and then made brittle, and its visible appearance is, of course, dental fluorosis. The enzyme poison effect extends to our genes; DNA cannot repair itself, and chromosomes are damaged. Work at the University of Missouri showed genital damage, targeting ovaries and testes. Also affected is inter uterine growth and development of the fetus, especially the nervous system. Increased incidence of Down's Syndrome has been documented. Fluorides are mutagenic. That is, they can cause the uncontrolled proliferation of cells we call cancer. This applies to cancer anywhere in the body; but bones are particularly picked out. The incidence of osteosarcoma in a study reporting in 1991 showed an unbelievable 50% increase. A report in 2005 in the New England Journal of Medicine showed a 400% increase in cancer of the thyroid in San Francisco during the period their water was fluoridated. My particular concern is the effect of fluorides on the thyroid gland. Perhaps I may remind you about thyroid disease. The thyroid gland produces hormones which control our metabolism—the rate at which we burn our fuel. Deficiency is relatively common, much more than is generally accepted by many medical authorities; a figure 1:4 or 1:3 by mid-life is more likely. The illness is insidious in its onset and progression. People become tired, cold, overweight, depressed, constipated; they suffer arthritis, hair loss, infertility, atherosclerosis, and chronic illness. Sadly, it is poorly diagnosed and poorly managed by very many doctors in this country.

What concerns me more deeply is that in concentrations as low as 1 ppm, fluorides damage the thyroid system on 4 different levels.

1) The enzyme manufacture of thyroid hormones within the thyroid gland itself. The process by which iodine is attached to the amino acid tyrosine and converted to the two significant thyroid hormones, thyroxine (T4) and triiodothyronine (T3) is slowed.
2) The stimulation of certain G proteins from the toxic effect of fluoride (whose function is to govern uptake of substances into each of the cells of the body) has the effect of switching off the uptake into the cell of the active thyroid hormone.
3) The thyroid control mechanism is compromised. The thyroid stimulating hormone output from the pituitary gland is inhibited by fluoride; thus reducing thyroid output of thyroid hormones.
4) Fluoride competes for the receptor sites on the thyroid gland which respond to the thyroid stimulating hormones; so that less of this hormone reaches the thyroid gland and so less thyroid hormones is manufactured.

These damaging effects, all of which occur with small concentrations of fluoride, have obvious and easily indentifiable effects on thyroid status. The running down of thyroid hormone means a slow slide into hypothyroidism. Already the incidence of hypothyroidism is increasing as a result of other environmental toxins and pollutions together with wide spread nutritional deficiencies.

141 million Europeans are at risk.

One further factor should give us deep anxiety. Professor Hume of Dandee, in his paper given earlier his year to the Novaris Foundation, pointed out that iodine deficiency is growing worldwide. There are 141 million Europeans at risk. Only five European countries are iodine sufficient. UK now falls into the marginal and focal category. Professor Hume recently produced figures to show that 40%

of pregnant women in the Tayside region of Scotland were deficient by at least half of the iodine required for a normal pregnancy. A relatively high level of missing, decayed, filled teeth was noted in this nonfluoridated area, suggesting that the iodine deficiency was causing early hypothyroidism, which interferes with the health of teeth. Dare one speculate on the result of now fluoridating the water?

"Displaces the iodine in body."

These figures would be worrying enough since they mean that iodine deficiency, which results in hypothyroidism (thyroid hormone cannot be manufactured without iodine), is likely to affect huge number of people. What makes it infinitely worse is that fluorine, being a halogen (chemically related to iodine but very much more active), displaces iodine. So that the uptake of iodine is compromised by the ejection, as it were, of the iodine by fluoride. To condemn the entire population already having marginal levels of iodine, to inevitable progressive failure of their thyroid system by fluoridating the water borders on criminal lunacy. I would like to place a scenario in front of those colleagues who favor fluoridation. A new pill is marketed. Some trials, not altogether satisfactory, nevertheless show a striking improvement in dental caries. Unfortunately, it has been found to be thyrotoxic, mutagenic, and immunosuppressive and causes arthritis and infertility in comparatively small doses over a relatively short period of time. Do you think it should be marketed?

Fluoridation of the nation's water supply will do little for our dental health but will have catastrophic effect on our general health. We cannot, must not, dare not, subject our nation to this appealing risk.

Or any nation worldwide.

Dr. Barry-Durrant-Peatfield obtained his medical degrees in 1960 at Guy's Hospital, London. He left the NHS in 1980 to specialize in the thyroid illnesses drawing inspiration from the work of infamous Dr. Broda Barnes, at the foundation that bears his name, Connecticut, USA. He has been a medical practitioner for over forty years specializing in metabolic disorders during which time he became a leading authority in the UK on thyroid and adrenal management. For over twenty years, he also ran a successful private clinic and became a nationwide leading authority on thyroid and adrenal dysfunctions but clashed with establishment medicine in the management of thyroid illness. He is the author of *The Great Thyroid Scandal.* He currently lectures at nutritional colleges in London as well as conducting his own teaching seminars. Dr. Barry-Durrant-Peatfield can be reached via email at info@drpeatfield.com.

The reference cited: L. Goldberg - La Semana Med. 28:628(1921) cited in Wilson RH, DeEdsF - "The synergistic Action of Fluoride Toxicity," *Endocrinology* 28:851 (1940).

http://www.rense.com/general57/FLUR.HTM

This speech or copy of this presentation should be provided to all our senators and congressman with a message "STOP POISONING OUR PEOPLE!" Additional reading material regarding the adverse effect of fluoride poisoning our thyroid gland can be accessed under the following links:

http://www.life-enthusiast.com/effects-of-fluoride-on-your-thyroid-a-4615...

http://www.thyroidnation.com/pineal-gland-thyroid-fluoride

http://www.fluoridealert.org/issues/health/thyroid

http://www.fluoridationqueensland.com/blog/2010/18/08/the-fluoride...

http://www.el-resource.org/articles/related.conditions-articles/...

http://www.articles.mercola.com/sites/.../08/13/fluoride-and-thyroid

There are other, too numerous articles regarding toxic fluoride effect on the thyroid gland to quote here. The reader can access these under *toxic fluoride effect on the thyroid gland on the Internet.*

Fluoride Effect on the Adrenal Gland

The adrenal glands are tiny organs located on top of each kidney. Adrenal glands secrete hormones that regulate chemical balance, regulate metabolism, and supplement other glands. The gland produces numerous hormones that control our development and growth, affect one's ability to deal with stress, and help to regulate the function of the kidneys. There are two parts to the adrenal gland—the cortex and the medulla, which produce hormones to regulate bodily functions. The medulla produces two hormones—norepinephrine (noradrenaline) and epinephrine (adrenaline). Adrenaline regulates stressful events. These hormones increase blood pressure, raise heartbeat, and cause sweating and increased breathing/respiration. Epinephrine and norepinephrine increase blood flow to muscles and brain, help with conversion of glycogen to glucose in the liver. Norepinephrine is also produced by the brain and is a very important neurotransmitter. It affects the operation of the human nervous system, vascular functions, liver processes, and mood regulation. During time of stress and anxiety it affects the processing of glucose, offers access to stored energy, provides increased oxygen to the brain, and increases blood flow to the muscles.

The cortex secretes several hormones that affect blood pressure, blood sugar levels, growth, and sexual characteristics. The cortex secretes three hormones—glucocorticoids, mineralocorticoids, and androgens. The glucocorticoid has three distinct hormones: Cortisol is released in response to stress to increase the glucose level in blood, metabolize fats, carbohydrates, and proteins, and suppress the immune system. The second, corticosterone, is produced as an intermediate element in the conversion of pregnenolone to aldosterone. The third, cortisone, suppresses the immune system. The mineralocorticoids consist of aldosterone, which is important in regulating blood pressure and salt and water balance and has a small effect on the metabolism of fats, carbohydrates, and proteins. Progesterone plays a role in embryogenesis, pregnancy, and menstrual cycle. Deoxycorticosterone is a precursor to aldosterone. Androgens are different hormones responsible for the development of male characteristics and secretion of testosterone.

http://www.fluoridealert.org/.../pesticides/effect.endocrine.adrenal.htm

http://www.metabolichealing.com/hypothyroidism.the-adrenals-the-intrinsic-link

Chapter 4

Dental Fluorosis

Dental fluorosis is a cosmetic condition that affects the teeth. It is caused by overexposure to fluoride during the first eight years of life. This is the time when most permanent teeth are formed. Teeth affected by fluorosis appear as mildly discolored. The teeth may have long white specs or other markings that only a dentist can detect. In more severe cases, the teeth may have stains, ranging from yellow to dark brown, surface irregularities, and pits in the enamel that are highly visible. http:www.webmd.com/fluorosis.html

It first attracted attention in the early twentieth century. Native-born residents of Colorado Springs had high incidents of so-called Colorado brown stain on their teeth. It was diagnosed of high fluoride content in their drinking water. These people with stains had a diminished number of cavities. This sparked the addition of fluoride to drinking water. This was done without any real study of the effect of fluoride on the rest of the body. Now, it has been determined that dental fluorosis affects one in four Americans, ages 6 to 49, prevalent mostly in the 12- to 15-year-old age group. Most cases are of the mild nature, but continued use of fluoridated water and fluoride from other sources can push this condition into the mild or severe condition. About 2% of the cases are of the moderate class and 1% are of the severe kind.

The major cause of dental fluorosis is the inappropriate use of fluoride containing dental products such as toothpaste, mouthwashes, and dental fluoride treatments. Children especially need to be watched that they do not swallow the toothpaste. Naturally occurring fluoride in water could be in the higher level than the recommended safe level. Initially it was established at 4 mg/liter, since revised to 0.7–1.2 mg/liter. No studies have ever been done to establish safe limits. Symptoms of dental fluorosis range from tiny white specks or streaks that may be unnoticeable to dark brown stains and rough enamel that is difficult to clean. Enamel fluorosis is the fluoride related alterations to enamel, which occur during development. In humans, plasma fluoride concentrations result from long-term ingestion of 1–10 ppm of fluoride in the drinking water. In severe cases, the enamel is porous, is poorly mineralized, stains brown, and contains relatively less mineral and more proteins than sound enamel. http://www.mouthandteeth.com/conditions.html

Dental Association finally established a classification for the different stages of dental fluorosis. The classification was split into several numerical classes according to the severity level. Clean teeth are rated as 0; mild level as 1–3; moderate level as 4–5; and severe level as 6–9. There are limited treatments for dental fluorosis. Bleaching is usually recommended for mild fluorosis (level 1–2). Treatment for moderate dental fluorosis includes micro-abrasion, where the outer affected area of enamel is abraded from the tooth surface in an acidic environment. Composite restoration combined with micro-abrasion or application of aesthetic veneers can be used at this level. Cases with severe fluorosis may necessitate prosthetic crowns.

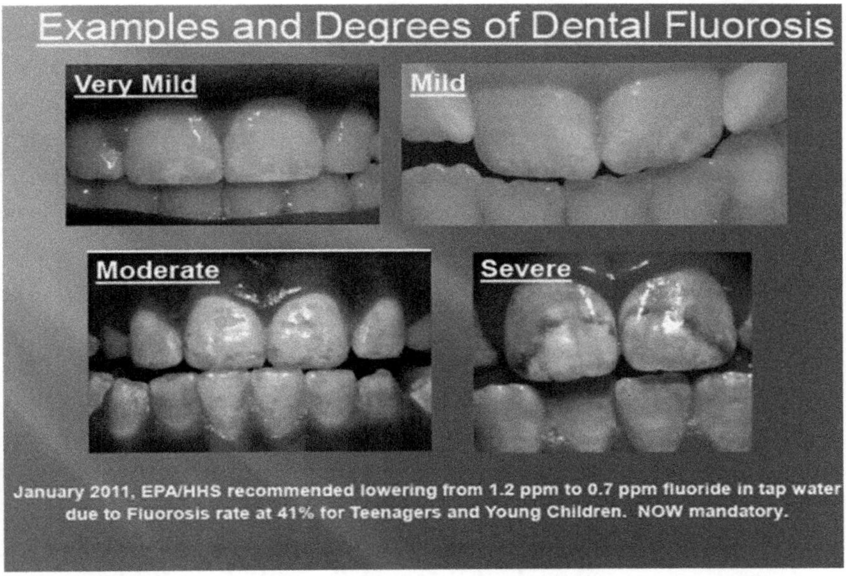

In the USA, approximately 10 million people are exposed to naturally occurring fluoride in the water. Areas that have extremely high levels of fluoride in the drinking water are as follows:

1) Colorado at 11.2 mg/liter
2) Oklahoma at 12.0 mg/liter
3) New Mexico at 13.0 mg/liter
4) Idaho at 15.9 mg/liter

The government propagated the following statements: "Fluoridation of public drinking water has significantly decreased the evidence of dental decay." **The only study that recommended the fluoridation of water was a project called Program F, which subsequently was established as the biggest fraud perpetrated on the citizens of the United States by the federal government. Where are the scientific studies to prove the benefits of fluoridation? There are none, but there are a large number of studies proving the detrimental effect on human health. Western Europe stopped fluoridating their drinking water some decades ago and still showed a decrease in dental cavities due to improved**

dental hygiene. Another misleading statement from the Centers for Disease Control and Prevention (CDC) that fluoridation of public drinking water was one of the great public health achievements of the twentieth century. How about the different health diseases it caused due to the poison fluoride? Tell it to parents with children with autism or ADD or ADHD. How about the elderly with an increase in bone fractures due to the brittle bones? How about obesity, which is on the increase due to malfunctioning thyroid glands? How about the increase in Alzheimer's disease due to pineal gland calcification? How about the increase in dental fluorosis? How about a decrease in fertility? Numerous scientific papers have established that fluoridating public drinking water is highly detrimental to public health. According to Dean Burg's statement it is **PUBLIC MURDER.**

What is really scary about fluoridation of public water system: the first intentional addition of fluoride in drinking water occurred in early 1930s in Nazi Germany. It was marketed as supposed to help children's teeth, but the actual sodium fluoride was to sterilize humans and force the masses into a calm docile state of submission and declining health state. This is one way to control the population through this manner. Research chemist Charles Perkins was sent by the US government to ascertain the truth of water fluoridation. His report was "the German chemists worked out a very ingenious and far reaching plan of mass control that was submitted to and accepted by the German General Staff. This plan was to control the population of any given area through mass medication of drinking water supplies. In this scheme of mass control, sodium fluoride occupied a prominent place. However, and I want to make this VERY DEFINITE, the real reason behind water fluoridation is not to benefit children's teeth. The real purpose behind water fluoridation is to REDUCE THE RESISTANCE OF THE MASSES to domination and control, and LOSS OF LIBERTY. Repeated doses of infinitesimal amounts of fluorine, within time, gradually reduce the individuals power to resist domination by slowly poisoning and narcotizing this area of the brain tissue, and make him submissive to

the will of those who wish to govern him. I was told of this entire scheme by a German chemist who was the official of the great Farben chemical industries and was prominent in the Nazi movement at the time. I say this with all the earnestness and sincerity of a scientist who has spent nearly 20 years researching in chemistry, biochemistry, physiology, and pathology of fluorine. Any person who drinks artificially fluoridated water for a period of one year or more, will never again be the same person, mentally or physically."

Is this what's happening to us now, to control us, with so many still advocating fluoridation?

http://www.greenfacts.org/en/fluoride/fluorosis-2/09-dental-fluorosis.htm

Chapter 5

Skeletal System

The skeletal system consists of about 206 bones and also includes a network of tendons, ligaments, and cartilage that connects them to each other or muscles. The vital function of the skeletal system is support, movement, blood cell production, and calcium storage. The skeleton of adult male and female are slightly different. This is to accommodate childbirth by the females. A typical bone has a very dense outer layer, next is a layer of spongy bone. In the middle of the bone is the bone marrow. The bone marrow is important since this is where new blood cells are produced. Teeth are considered part of the skeletal system. The teeth are made up of dentin and enamel, which is the strongest skeletal structure in our body.

The skeletal system is composed of two distinct section. The first is the axial skeleton, consisting of 80 bones, which is the vertebral column, rib cage, and the skull. The axial skeleton transmits the weight from head, trunk, and upper extremities down to the lower extremities at the hip joints, which help humans to maintain an upright posture. The second part consists of appendicular skeleton, containing 125 bones. It is formed by the pectoral girdles, upper limbs, pelvic girdle, and lower limbs. Their main function is to facilitate walking, running, and other movements. They also protect the major organs. Diseases of the skeletal system consist of osteoporosis and arthritis, which is a group of quite a few inflammatory diseases that damage joints and their surrounding structure. It usually affects the joints of the neck, shoulders, hands, lower back, hips, and knees. Leukemia is a cancer of the blood. Skeletal system is involved since cancer starts in the marrow. Bursitis is a disorder that causes pain in the body's joints. It can affect the shoulder and hip joints caused by an inflammation of the bursa. These are small fluid-filled bags that act as a lubricating surfaces for muscles to move over bones. The skeletal system is susceptible to breaks, strains, sprains, and fractures.

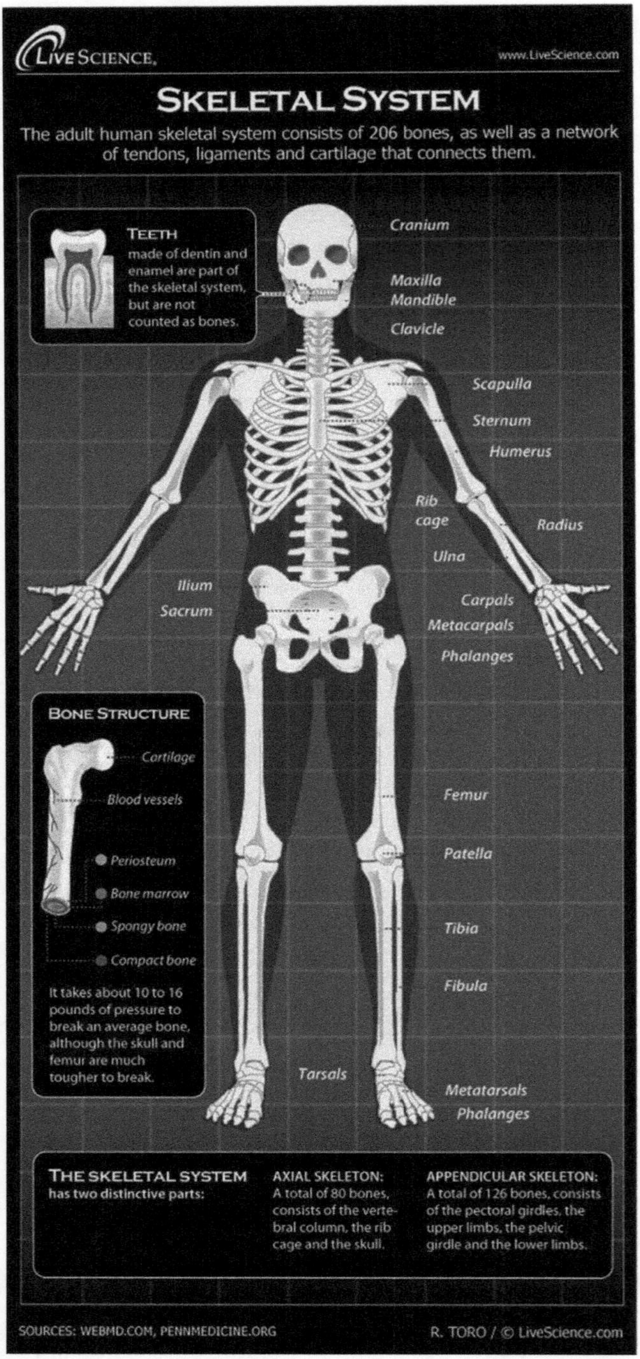
LIVE SCIENCE
www.LiveScience.com
SKELETAL SYSTEM
The adult human skeletal system consists of 206 bones, as well as a network of tendons, ligaments and cartilage that connects them.
TEETH
made of dentin and enamel are part of the skeletal system, but are not counted as bones.
Cranium
Maxilla
Mandible
Clavicle
Scapulla
Sternum
Humerus
Rib cage
Radius
Ulna
Ilium
Sacrum
Carpals
Metacarpals
Phalanges
BONE STRUCTURE
Cartilage
Blood vessels
Periosteum
Bone marrow
Spongy bone
Compact bone
It takes about 10 to 16 pounds of pressure to break an average bone, although the skull and femur are much tougher to break.
Femur
Patella
Tibia
Fibula
Tarsals
Metatarsals
Phalanges
THE SKELETAL SYSTEM has two distinctive parts:
AXIAL SKELETON: A total of 80 bones, consists of the vertebral column, the rib cage and the skull.
APPENDICULAR SKELETON: A total of 126 bones, consists of the pectoral girdles, the upper limbs, the pelvic girdle and the lower limbs.
SOURCES: WEBMD.COM, PENNMEDICINE.ORG
R. TORO / © LiveScience.com

Skeletal Fluorosis

A second sign of increased fluoride intake is skeletal fluorosis. The disease is also known as crippling fluorosis. It develops in stages. It is not easily recognized and can easily be misdiagnosed. Skeletal fluorosis is a bone disease caused by excessive consumption of fluoride, as evidenced from fluoride endemic drinking water. In advanced cases, skeletal fluorosis causes pain and damage to bones and joints. First signs of dental fluorosis can be a signal for skeletal fluorosis, since this is due to increased fluoride intake. According to the US government experts, most people currently ingest about four times as much fluoride as they did during the early days of water fluoridation—approximately equally divided between drinking water, food, other beverages, fluoride in toothpaste, mouth rinses, drops, tablets, vitamins, gels, sealants, and fluoridated milk. Fluoride is also found in medicines, anesthetics, pesticides, herbicides, uranium, and many other items. They are released into the air, water, and soil, constantly increasing our exposure to this bioaccumulative substance, which is more toxic than lead. Common causes of fluorosis include inhalation of fluoride dusts and/or fumes by workers in industry, use of coal as an indoor fuel source, and consumption of dark tea, particularly brick tea. It can also be caused by fluoridated fumigants like cryolite and surfuryl. (More on these two compounds later.)

The diagnosis of skeletal fluorosis, until recent years, was made with the help of radiographs that reveal interosseous membrane calcification, increased bone density, and bone mass. These are late characteristics of the disease. Recognizing the disease at such late stages does not help prevention. The disease is usually irreversible by then.

In some areas, dental and skeletal fluorosis is endemic, where fluoride is already in high concentrations in local drinking water. In some states such as Colorado, Oklahoma, New Mexico, and Idaho. You should be able to find out your water's fluoride content by getting the information from your local water department.

Skeletal fluorosis manifests itself in several phases. In the preliminary phase, the symptoms are increases in bone mass, which is slightly detectible through radiography. In the clinical phase 1, the symptoms are sporadic pain, stiffness of joints, and osteoporosis of pelvis and vertebral spine. In the clinical phase 2, the symptoms are chronic joint pain, arthritic symptoms, slight calcification of ligaments, and increased osteoporosis and cancellous bones, with or without osteoporosis of long bones. In the clinical phase 3, the symptoms are limitation of joint movement, calcification of ligaments of each vertebral column, crippling deformities of the spine and major joints, muscle wasting, neurological defects, and compression of the spinal cord. The symptoms are mainly promoted in the bone structure. Due to high fluoride concentration in the body, the bones are hardened and thus less elastic, resulting in an increased frequency of fractures. The bone fractures due to brittle bones have steadily increased over the past decades. Could this be true due to advent of fluoridation of water?

http://www.consumerhealth.org/articles/display.ctm?ID=1999030322283

Other symptoms include thickening of the bone structure and accumulation of bone tissue, which contribute to poor joint mobility. Ligaments and cartilage can become ossified. Some patients have side effects such as ruptures of the stomach lining and nausea. Fluoride damages the thyroid gland leading to hyperparathyroidism, the uncontrolled secretion of parathyroid hormones. Parathyroid glands are the four bean-sized glands in the neck by the thyroid gland. These hormones regulate calcium concentration in the body. An elevated parathyroid hormone results in a depletion of calcium in bone structures and thus a higher concentration in blood. Bone flexibility is diminished resulting in the bone more prone for fractures.

Any sign of dental fluorosis is a good indication that one is exposed to increased concentration to fluoride, which could also signal potential effect on the bones. It would be highly beneficial for anyone having been diagnosed with dental fluorosis, to follow this with a radiography to determine any potential effect on the bones. Metabolic bone disease occurs more frequently in endemic fluoride areas than in residents of non-endemic areas. Common metabolic bone diseases associated with endemic skeletal fluorosis were osteoporosis (bone resorption), rickets, osteomalacia, and parathyroid bone disease. Fluoride, by its nature of its incorporation into bone crystals and by its direct cytotoxic effect on the bone resorbing cells, reduces the availability of calcium from the bone.

The mechanism leading to the hyperparathyroidism is not quite clear. Further studies are needed. One thing is clear—skeletal fluorosis increased in high fluoride concentrations in fluoride endemic areas.

As a teenager, I was following professional sports like baseball, football, hockey and basketball. Regarding baseball, during the '50s and '60s, maybe even '70s, the so-called Tommy John surgery was hardly mentioned. Lately, this operation on the pitchers, I would say, has reached epidemic proportions. Could this be due to the fluoride effect of the calcification of the ligaments? How about the increase in head concussions due to fluoride effect on the brain? It takes the athletes a long time to recover from ankle, knee, or other sprains. Could this also be a contributing factor from excess fluoride? Some of these questions need to be answered since it could be due to some form of skeletal fluorosis.

http://www.nealthhype.com;ABCD-FirstAid.Injuries
http://www.en.wikipedia.org/wiki/Skeletal_fluorosis
http://www.fluoridealert.org/issues/health/skeletal_fluorosis

Chapter 6

Fluoride and the Field of Athletics

I started to follow professional sports teams sometime during the year 1952. I always enjoyed sports. My favorite sport was soccer since my first fifteen years were spent in Europe. Soccer was not practiced or known in the US until much later in my life. Since its introduction, quite a few soccer players have become proficient field goal kickers in the NFL.

I remember following the Detroit Tigers, Detroit Lions, and Detroit Red Wings. Following the Detroit Pistons came much later. One of the things that impressed me was the pitchers who could pitch complete games. One of the biggest prides of those pitchers at that time frame was the number of complete games statistic. There were no instances of so-called tired arms, Tommy John surgeries to repair pitching arms, or other muscular strains, sprains to such an extent as it is today. At the most, there were only so-called tennis elbows and some sprains and strains. Presently, a season doesn't occur without a number of Tommy John surgeries or ACL repair procedures at both college and professional level. It is also creeping in at the high school level.

Having accessed many studies regarding fluoride poisoning, what caught my attention was reading about the calcification of ligaments. There is ample evidence that fluoride disrupts collagen formation, which affects skin wrinkles, ligaments, tendons, and cartilage (*The Aging Factor* by Dr. John Yiamouyeannis).

Tendons have been misunderstood for quite some time. The general belief is that some every day minor aches and pains are as a result of aging. But what about the young sport players who exercised strenuously for their respective sport of choice. In my time the aches and pains were due to exertion, disappearing upon rest.

Tendons and ligaments are tough, flexible bands of mostly collagen that connect our muscles to the bones. They are like internal rubber bands to keep our muscles and bones in tune with our movements. That also means that they can only be stretched so far and/or so many times before they may start to signal a sprain, strain, or other muscular discomfort. As an athlete, you understand the exercises you do to condition these ligaments/tendons to repetitive physical activity in whatever sport of your choice. Over time these tendons take quite a beating. Every move we make, there are hundreds of different tendons activated.

One of the problems is that tendons are hard to nourish because they belong to the avascular system—meaning that they have a limited blood supply and in some spots none at all. To make matters worse, fluoride that is in our drinking water and sports drinks based on fluoridated water base are very detrimental to the health of our collagen that is in our tendons, ligaments, and cartilage.

Disrupted collagen formation leads to calcification of tissues that should not be calcified. That leads to stiff and possibly torn ligaments or tendons, muscle aches, cartilage decline, arthritis, weakened arteries, brittle bones, and in some instances bone spurs. Could this be the cause of so many Tommy John surgeries since the fluoride adversely affects calcification of ligaments and tendons. Bone spurs that developed as a result of fluoride presence in our water or sport drinks have to be surgically removed. How about all the ACL damages to the athletes in their elbows, wrists, and knees, especially those playing a sport like football? How about brain concussions?

Chinese studies have shown the damage to the brain cells as early as a fetus in a mother's womb. So do the neurotoxins fluoride, aspartame, and MSG (monosodium glutamate) cause an increase in brain concussions due to the damaged brain cells in the womb in early life, since the mothers drank fluoridated water or used diet drinks sweetened with aspartame or ate food containing MSG? How about the numerous vaccinations received at an early stage of life that contain numerous toxic adjuvant material? (Separate chapters cover these neurotoxins and their effect on human health in this book.) There is also an increase in sprained, twisted ankles and wrists, and sprained knees as well. How about bad backs and sprained muscles? Some of the sprained ankles or wrists or knees can linger for quite some time, especially if blood vessel is ruptured in that area and spilled internal blood is not removed, since it is in an area of limited lymphatic circulation. I experienced such a nasty ankle sprain. When it happened, I could hear the bones crack.

This happened to me while I was working around the house patching all the cracks with caulk. I had to use a ladder, and while stepping off of it, I sprained my ankle. This was on a Saturday, and I had quite a bit to do yet to finish the job. My mother some time ago made an ointment of natural compounds, which was good to alleviate bruises. I applied this on my ankle, had dinner, took a bath, reapplied the ointment, and went to bed. The next morning, I got up and went to finish the job of caulking the house. I completely forgot my sprained ankle until I stepped down from the ladder. I felt a slight pain deep inside my ankle. That's when it became apparent to me that I sprained my ankle, but is this ointment good. That was not the end of it. Two years later, I was stung by a yellow jacket on the instep on the leg that I sprained my ankle previously. I applied the ointment to alleviate the pain, which it did within three minutes. We were leaving the next afternoon on a trip to Florida from Troy, Michigan. At the first stop that day, I pulled the sock off, and to my surprise, I had a large dollar-size black-and-blue area where I was stung by the yellow jacket. I applied the ointment with the thought that if it did not help I would have to seek medical help. The next day when I looked at the area, it was still black and blue but less intense in color. I reapplied the ointment and decided to check it at lunchtime. When I did that, the black-and-blue area completely disappeared. What really surprised me that there was a rectangular black area on the instep the size of an elongated centimeter. The only explanation I can come up with is that the blood from the sprained ankle was finally being drawn out of the sprained area. This is two years later though.

This makes me wonder, that if that is true, since athletes with sprained ankles take quite a lengthy time to heal, would something like this ointment, applied immediately after a sprain diminish the lost playing time? I don't know. This I do know—since I tried to make this ointment myself and while mixing the ingredients in a container at the boiling point of water, I spilled some of the ingredients on my hands and rubbed the stuff around. Sometime later, my wrists started aching where I could feel every heartbeat. It lasted a day and a half. Since then I have nice, pain-free wrist mobility, pain-free for the last twenty-five years and counting.

But back to the professional athletes. What the athletic departments should do is question the athletes as to their drinking habits regarding soda drinks, juices, and sports drinks, such as Gatorade. Gatorade used to be all-natural until it was sold and became too commercialized. One of the biggest culprits is diet soda, which probably contained fluoridated water to start with. Gatorade, which probably also is made with fluoridated water, may contain artificial colors and sweeteners such as aspartame or high-fructose corn syrup made from a GMO corn, which are just as damaging to one's health. In the professional sports, the athletes sign long multimillion-dollar contracts without realizing that their lifestyle as to what they eat and drink could shorten their productive years as an athlete. There are numerous baseball/football/hockey/basketball examples of successful athletes for a few years fading very fast sometimes without realizing as to their cause of effectiveness on the field, arena, or court. I hope that this information in this book can be a wake-up call to the participants in sport. You don't have to take my word for it, but access some of the Internet links provided to get yourself informed regarding the abovementioned toxins. I would not trust some of the so-called authorities, since one never knows who they align themselves with. They usually align themselves with the money and propagate the misinformation taught in schools.

One of the easiest ways to provide safe drinking water or soda is to install reverse osmosis systems in the clubhouses. To have carbonated water—use the soda stream system. As to flavors, limit to all verified natural ones.

I hope this gives some athletes a little insight and warning if they want to last a number of years in their respective chosen athletic endeavor. If you want organic food, buy it from the Amish. They are all natural as nature intended it to be. Better yet, please investigate all the Internet links provided to educate yourself as to your health and the well-being of your family. Regardless of what high-level professionals state regarding our environment and health, uncontaminated food supply is the best medicine. It is not the processed food, aspartame-containing staples, GMO crops, and numerous other foods containing a number of artificial ingredients and toxic chemicals in our drinks and food that permeate in our society.

Chapter 7

Artificial Sweetener Aspartame

Aspartame is a manmade artificial sweetener that is about two hundred times more sweet than sugar. It is composed of two amino acids—phenylalanine and aspartic acid. It is held together in a methyl ester bond 10% of it is contained in aspartame. It is best known by its commercial names as NutraSweet, Equal, Sweet One, and Spoonful. It is a synthetic combination of 50% phenylalanine, 40% of aspartic acid, and 10% of methanol (wood alcohol).

Process for producing aspartame. European Patent Application EU0036258.

http://www.freepatentonline.com/EU0036258.html

The artificial sweetener aspartame, a dipeptide with the formula Asp-Phe-Me, is produced using a cloned microorganism. A DNA, which codes for a large stable peptide comprised of the repeating amino acid sequence (Asp-Phe), is inserted into a cloning vehicle, which in turn is introduced into a suitable host microorganism. The host microorganism is cultured and the large peptide containing the repeating Asp-Phe sequence is harvested therefrom. The free carboxyl group of the large peptide is benzylated and then hydrolyzed to benzyl Asp-Phe dipeptides. This dipeptide is methylated and then debenzylated to form aspartame.

http://www.independent.co.uk./news/worlds-top-sweetener-is-made-with-gm-bacteria-1101176.html

Aspartame is added to thousands of food products, dairy products, drinks, candy, gum, vitamins, health supplements, and even pharmaceuticals. Each of the three compounds in aspartame poses its own danger and is well documented. Aspartame at temperature of 86 degrees Fahrenheit breaks down to phenylalanine, aspartic acid, methanol, and trace amounts of a compound called diketopiperazine (DKP). So far very little is known about DKP compound's effect on the human body. We need studies to determine what effect it has if any on the human body. Let's take a look at each compound that aspartame is made of, and its effect on the human psychology and physiology.

Phenylalanine

Phenylalanine is an essential amino acid found in many proteins. Major sources of phenylalanine are proteins in meat, fish, eggs, cheese, and milk. There are two forms of phenylalanine. D-phenylalanine is a mirror image of the L-phenylalanine, but of the two forms, the L-form is the essential amino acid and is the only form found in food sources. Phenylalanine is converted to the amino acid tyrosine, which is important in the production of many neurotransmitters, such as norepinephrine (noradrenaline), epinephrine (adrenaline), and dopamine. It is also important in thyroid gland hormone production, which controls the metabolism of the body. Norepinephrine is important for concentration, motivation, and alertness. It travels through the blood system to stimulate the brain. Epinephrine is produced in times of stress or danger, to ensure proper supply of oxygen. Dopamine helps in staying motivated and assertive and helps in handling daily stress and problems. Too little of the amino acid affects the above neurotransmitters and the adverse effect it has on the mental and body functions covered in previous chapter. One would think that aspartame contains this essential amino acid would be a good thing, but too much of a good thing can and is damaging to one's health.

Too much phenylalanine has an adverse effect. One is that some people are born with a condition called phenylketonuria (PKU).

Newborn babies are supposed to be tested for this condition within twenty-four to forty-eight hours of birth. What that means, these people do not have the necessary enzyme to metabolize the phenylalanine to usable form. The amount of the amino acid in the blood builds up to toxic levels, ultimately causing severe brain disorders, including mental retardation and seizures. These people are warned not to eat foods containing phenylalanine, and as such to stay away from foods and drinks containing aspartame. On ingestion, it forms the three compounds it is formed from—50% phenylalanine, 40% aspartic acid, and 10% methyl alcohol. The increased level of phenylalanine has the following adverse effects—it may affect blood pressure and cause headaches, nausea, and heartburn. Ingesting of phenylalanine contained in aspartame raises a high level of phenylalanine in the hypothalamus, medulla oblongata, and corpus striatum in areas of the brain according to numerous investigative studies. Dr. Russell Blaylock, a neurosurgeon, points out that excessive buildup of phenylalanine in the brain can cause schizophrenia or seizures.

(Ref.: *Journal of Neurochemistry*, volume 17, issue 4, p. 469–474, April 1970)

Aspartic acid

The second ingredient in aspartame. Aspartic acid acts as a neurotransmitter in the brain by facilitating the transmission of information between neurons. Again, too much aspartate in the brain kills certain neurons by allowing too much influx of calcium into the cells. This influx of calcium triggers excessive amounts of free radicals, which kill the cell. This effect on the neuron cells by aspartate classifies aspartate as a neurotoxin. The blood-brain barrier protects the brain from excess aspartate, as well as other toxins. Causes for aspartate to pass into the brain could be due to the following—the blood-brain barrier is (1) not fully developed during childhood, (2) does not fully protect all areas of the brain, (3) is damaged by numerous chronic and acute conditions, or (4) allows seepage of excess aspartate into the brain even when intact. (Could this be due to the fluoride effect on brain tissue?) The excess aspartate begins slowly to destroy the neurons in the brain by a large number, before any symptoms of the chronic illness manifests itself. Such chronic illnesses that have been shown to be contributed by long-term exposure to excitatory amino acids such as is in aspartame. These can be or may be MS (multiple sclerosis), Alzheimer's disease, Parkinson's disease, memory loss, hormonal loss with their own adverse multiple illnesses, neuroendocrine disorders, dementia, hypoglycemia, etc. Aspartic acid has a damaging effect on the body when it is isolated from its naturally protein-bound state and causes it to become neurotoxin

Methyl alcohol

The third major ingredient that is released when aspartame is ingested is methyl alcohol. Methanol is also known as wood alcohol. Methanol is released within hours after ingestion of aspartame by hydrolysis of the methyl group. Methanol is a toxin that destroys the myelin tissue in the body. Myelin is an insulating layer that forms around neurons, including those in the brain and spinal cord. It is made up of protein and fatty substances. The purpose of myelin sheath is to allow impulses to transmit quickly and efficiently along the nerve cells. If myelin is damaged, the impulses slow down. It can cause or mimic a disease such as multiple sclerosis, migraine headaches, affect vision.

The main reason for methanol toxicity is that it breaks down in the body into formaldehyde and formic acid. Formaldehyde is a neurotoxin and classified as a carcinogen. Even EPA (Environmental Protection Agency) states, **"That methanol is considered a cumulative poison due to the low rate of excretion once it is absorbed in the body. Methanol is oxidized to formaldehyde and formic acid, both of these metabolites are toxic."** The recommended dose by EPA is 7.8 mg/day.

Dr. Woodrow Monte, PhD, RD, director of the Food Science and Nutrition Laboratory at Arizona State University, states, **"When diet sodas and soft drinks, sweetened with aspartame are used to replace fluid lows during exercise and physical exertion in hot climates, the intake of methanol can exceed 250 mg/day or 32 times**

the EPA's recommended limit of consumption for this cumulative toxin."

Symptoms from methanol poisoning are headaches, ear buzzing, dizziness, nausea, gastrointestinal disturbances, weakness, vertigo, chills, memory lapses, numbness, shooting pains in the extremities, behavioral disturbances, and neuritis. Most notable methanol poisoning affect vision, progressive contraction of visual field, blurring of vision, obscuration of vision, retinal damage, and ultimately blindness. Formaldehyde causes retinal damage, interferes with DNA replication, and may cause birth defects. A study on rats indicates a chronic effect of aspartame on oxidative stress in the brain, with significant increase in lipid peroxidation, and superoxide dismutase activity. Chronic exposure of aspartame results in detectable methanol in the blood resulting in the oxidative stress in the brain. Oxidative stress can be defined as the state in which damaging free radicals outnumber one's antioxidant defenses, which lead to accelerated tissue and organ damage. Aspartame leads to decreased amount of glutathione, the active antioxidant form of glutathione, and reduced glutathione reductase activity.

A new review by scientists from the University of Pretoria and the University of Limpopo found that high doses of the sweetener aspartame may lead to neurodegeneration. It was also previously found that aspartame consumption levels can cause neurological and behavioral disturbances in sensitive individuals.

The review found that a number of direct and indirect changes occur in the brain as a result of aspartame consumption:

* The metabolism of amino acids
* Protein structure and metabolism
* The integrity of nucleic acids
* Neuronal function
* Endocrine balances

The breakdown of aspartame causes nerves to fire excessively, which indirectly leads to a high rate of neuron depolarization. These are growing concerns but conveniently ignored by the FDA and the EFSA (European Food Safety Authority).

History of Aspartame

Aspartame was accidentally discovered by a scientist working for G. D. Searle, involved in trying to formulate a new ulcer drug. It was accidentally spilled and ultimately exhibited a sweet smell. When tasted, it was determined that it was about two hundred times sweeter than sugar. This was in December 1965.

The following is a brief synopsis of G. D. Searle's request to approve aspartame as a food additive. In the spring of 1967, G. D. Searle is doing safety tests on aspartame prior to submission to the FDA. In the fall of 1967, Searle hired Dr. Harold Waisman, biochemist, to conduct safety tests on monkeys. Of the seven monkeys, one died and five came down with grand mal seizure. (Grand mal seizures are seizures that involve muscle contractions, muscle rigidity, and loss of consciousness. These seizures result from abnormal electrical activity in the brain.) In 1971, Dr. Olney informs Searle that aspartic acid caused holes in mice's brain, which was confirmed by Searle's scientists. In February 1973, Searle applied for FDA approval and submitted about one hundred studies claiming aspartame's safety. On March 5, 1973, FDA scientists stated that the safety information submitted by Searle was insufficient. They recommended further studies to be done. In July 1974, FDA granted aspartame its first approval for restricted use in dry foods. In August 1974, Attorney Jim Turner and Dr. John Olney filed first objections against aspartame approval. In March 1976, Turner's and Olney's petition triggered FDA investigation of laboratory practices by Searle. Results found practices to be shoddy. Investigators report stated that they **"had never seen anything as bad as Searle's testing."**

In January 10, 1977, FDA formally requested US Attorney's Office in Chicago to begin grand jury proceedings to investigate whether indictments should be filled against Searle, for knowingly misrepresenting findings and concealing material facts and false statements. This was the first time that FDA requested criminal investigation of a manufacturer. In March 8, 1977, G. D. Searle hired Donald Rumsfeld as the new CEO. In August 1, 1977, the Bressler Report, compiled by FDA investigators, was released. The report found that 98 of 196 animals died during Searle's studies. In December 1977, the US attorney who was assigned the investigation of Searle resigns accepted a position with Searle's law firm. In June 1, 1979, FDA established the Public Board of Inquiry to rule on safety issues surrounding now so-called NeutraSweet. In September 30, 1980, the Public Board of Inquiry concluded NeutraSweet should not be approved, pending further investigation of brain tumors. In January 1981, Donald Rumsfeld stated during a board meeting that he would push to get approval for aspartame (NeutraSweet) . In January 21, 1981, Donald Rumsfeld was sworn in as part of President Ronald Reagan's transition team. In March 1981, the FDA commissioners' panel established to review issues raised by Public Board of Inquiry. In July 15, 1981, one of the first acts of the new FDA commissioner Dr. Arthur Hayes was to overrule the Public Board of Inquiry, ignore the recommendations of his own internal FDA team, and approve NeutraSweet for dry foods again. It was previously rescinded, pending ongoing investigations. In October 15, 1982, FDA announced that Searle filed a petition to approve aspartame as a sweetener for carbonated beverages. In July 1, 1983, the NSDA (National Soft Drink Association) drafted an objection to the final ruling. The association claimed that Searle had not provided responsible certainty that aspartame and its degradation products were safe to use in soft drinks. In August 8, 1983, consumer attorney Jim Turner and Dr. Woodrow Monte filed a lawsuit with FDA objecting to aspartame approval based on unresolved safety issues. In September 1983, FDA Commissioner Dr. Arthur Hayes resigned under cloud of controversy and was hired by Searle public relations firm. In the fall of 1983 first carbonated drinks containing aspartame were released to the marked. In November 1984, Centers for Disease Control released "Evaluation of Consumer Complaints related to Aspartame (summary by Mullarkey). On November 3, 1987, US Senate hearings on "NeutraSweet Health and Safety Concerns" by the Committee on Labor and Human Resources by Senator Howard Metzenbaum began. For full detailed and expanded chronological order of aspartame history approval process, go to the following:

(http://www.dorway.com/history-of-aspartame)

(http://www.freezerbox.com/archive/2001/04/biotech/)

Aspartame was denied approval by FDA on three numerous occasions between 1973 and 1975.

(http://www.qualityassurance.synthasite.com)

Reported Aspartame Toxicity Effects

Toxicity reactions to aspartame ingestion can be divided into three distinct sections.

1) Acute toxicity reaction occurring within forty-eight hours of ingestion of an aspartame containing product
2) Chronic toxicity effects occurring anywhere from several days of use to appearing a number of years (e.g., one to twenty years) after the beginning of aspartame use
3) Potential toxicity effects that would be nearly impossible for the user to recognize the link to aspartame

Following is a listing from 661 persons who reported toxic effects from aspartame ingestion. Effects listed by different body parts.

	#people	%
Eye		
Decreased vision and/or other eye problems (blurring, bright flashes, tunnel vision)	140	25%
Pain for or both eyes	51	9%
Decreased tears, trouble with contact lenses or both	46	8%
Blindness (one or both eyes)	14	3%
Ear		
Tinnitus (ringing or buzzing sensation)	73	13%
Severe intolerance for noise	47	9%
Marked impairment of hearing	25	5%
Neurologic		
Headaches	249	45%
Dizziness, unsteadiness, or both	217	39%
Confusion, memory loss, or both	157	29%
Severe drowsiness and sleepiness	93	17%
Paresthesias (pins and needles or tingling) or numbness	82	15%
Convulsions (grand mal epileptic attacks)	80	15%
Petit mal attacks and "absences"	18	3%
Severe slurring of speech	64	12%
Severe tremors	51	9%
Severe hyperactivity and "restless legs"	43	8%
Atypical facial pain	38	7%

Psychological/psychiatric		
Severe depression	139	25%
Extreme irritability	125	23%
Severe anxiety attacks	105	19%
Marked personality changes	88	16%
Recent severe insomnia	76	15%
Severe aggravation of phobias	41	7%

Chest		
Palpitations, tachycardia (rapid heart action) or both	88	16%
Shortness of breath	54	10%
Atypical chest pain	44	8%
Recent hypertension (high blood pressure)	34	6%

Gastrointestinal		
Nausea	79	14%
Diarrhea (associated gross blood in the stool)	70	13%
Abdominal pain	70	13%
Pain on swallowing	28	5%

Skin and allergies		
Severe itching without a rash	44	8%
Severe lip and mouth reactions	29	5%
Urticaria (hives)	25	5%
Other eruptions	48	9%
Aggravation of respiratory allergies	10	2%

Endocrine and metabolic		
Problem with diabetes, loss of control, precipitation of clinical diabetes, or aggravation or simulation of diabetic complications	60	11%
Menstrual changes (severe reduction or cessation of periods)	45	6%
Paradoxical weight gain	34	6%
Marked weight loss	26	5%
Marked thinning or loss of the hair	32	6%
Aggravated hypoglycemia (low blood sugar attacks)	25	5%
Other		
Frequency of voiding (day and night), burning on urination (dysuria), or both	69	13%
Excessive thirst	65	12%
Severe joint pain	58	11%
Bloat	57	11%
Fluid retention and leg swelling	20	4%
Increased susceptibility to infection	7	1%

Interesting to note that on April 20, 1995, the Department of Health and Human Services released a list of symptoms attributed to aspartame in complaints submitted to the FDA.

Department of Health and Human Services Reported Symptoms

Reported symptoms	Number of complaints	Percent of reports	Percent of complaints
Headache	1847	21.10%	19.00%
Dizziness/poor equilibrium	725	11.20%	7.50%
Change of mood	656	10.00%	6.70%
Vomiting or nausea	647	9.80%	6.60%
Abdominal pain and cramps	483	6.90%	4.70%
Change in vision	362	5.50%	3.70%
Diarrhea	330	5.00%	3.40%
Seizures and convulsions	290	4.40%	3.00%
Memory loss	255	3.90%	2.60%
Fatigue, weakness	242	3.70%	2.50%
Other neurological symptoms	230	3.50%	2.40%
Rash	226	3.40%	2.40%
Sleep problems	201	3.10%	2.10%
Hives	191	2.90%	2.00%
Change in heart rate	185	2.80%	1.90%
Itching	175	2.70%	1.80%
Grand mal	174	2.60%	1.80%
Numbness, tingling	172	2.60%	1.80%
Local swelling	114	1.70%	1.20%
Change in activity level	113	1.70%	1.20%

Difficulty in breathing	112	1.70%	1.20%
Oral sensory changes	108	1.60%	1.10%
Change in menstrual pattern	107	1.60%	1.10%
Symptoms reported less than 100	1812	–	18.60%

Distribution of reactions to aspartame by product name

Reported symptom	Number of complaints	Percent of reports	Percent of complaints
Diet soft drinks	3021	45.00%	38.30%
Table top sweeteners	1714	26.10%	21.70%
Puddings, gelatins	623	9.60%	8.00%
Lemonade	410	6.20%	5.20%
Other	346	5.30%	4.40%
Kool-Aid	339	5.10%	4.30%
Iced tea	319	4.80%	4.00%
Chewing gum	319	4.80%	4.00%
Hot chocolate	318	4.80%	4.00%
Frozen confections	136	2.10%	1.70%
Cereal	119	1.80%	1.50%
Sugar substitute tablets	71	1.10%	0.90%
Breath mints	62	0.90%	0.80%
Punch mix	45	0.70%	0.60%

Fruit drinks	24	0.40%	0.30%
Nondairy toppings	8	0.10%	0.10%
Chewable multivitamins	8	0.10%	0.10%
Fruit, dried	1	0.01%	0.01%

Following articles reprinted (with permission) in its entirety on the toxic nature of aspartame present in numerous foods, soft drinks, and other food products.

Aspartame and Psychiatric Disorders

By Ralph Walton, MD
9-20-7

> Although psychiatry is far from an exact science, over the past half century there has been an explosive growth in our understanding of the human brain and consequently in our ability to diagnose and treat mental disorders. We have also become much more sophisticated about the impact of a variety of toxins on psychological processes. I am convinced that one such toxin is aspartame.
>
> Two years after aspartame was introduced onto the market I first became aware of the negative impact of this artificial sweetener on the central nervous system. I had been treating then a 54-year old woman with migraine, a tricyclic antidepressant, because of recurrent major depressive episodes. Previous psychoanalytically based therapy had proven ineffective, but she responded dramatically to 150 mg of imipramine per day. She had done well for 11 years on this medication, but

was then suddenly hospitalized with a grand-mal seizure and subsequent manic episode. One could postulate that she was bipolar, and the antidepressant had triggered the mania—but she had been on the same medication for a total of 11 years, and for the previous 5 years at the same 150 mg per day dose. Neither the seizure nor her mania was consistent with what we know about the clinical course of bipolar disorder or epilepsy. Careful history revealed that the only change in her life was a recent decision to switch from the sugar which she had always used to sweeten her iced tea to a newly marketed product with aspartame. Since aspartame can alter the balance of certain neurotransmitters which we believe are involved in mood disorders and can, in my opinion, alter the seizure threshold, I advised my patient to avoid all aspartame products. She did so, and had no further seizures, no further manic or depressive episodes. I discontinued the lithium carbonate which I had started when I mistakenly concluded that she had a bipolar disorder, reinstated her imipramine and she has continued to do well. After this case report was published in the medical literature, many patients with unexplained seizures or treatment resistant psychiatric problems were referred to me. I became increasingly convinced that aspartame could both trigger seizure activity and mimic or exacerbate a variety of psychiatric disorders. I presented a paper based on those patients at a 1987 MIT sponsored conference on Dietary

Phenylalanine and Brain Function

Industry sponsored criticism was made that my conclusions regarding aspartame's toxicity could not be accepted as valid because my case reports were "merely anecdotal" and not based on double blind research. Unfortunately, case reports do not currently have the respect in the mainstream medical literature which they deserve (historically much of medical progress has been based on careful observation of individual patients).

Nevertheless, I was so convinced of aspartame's toxicity, and the need to have its hazards more widely appreciated in the medical community, that I did undertake a double blind study. That study "Adverse Reactions to Aspartame: Double-Blind Challenge in Patients from a Vulnerable Population" was published in Biological Psychiatry in 1993. It demonstrated that individuals with mood disorders are particularly sensitive to aspartame and experienced an accentuation of depression and multiple physical symptoms. I had expected that the difficulties experienced by patients receiving aspartame would be fairly subtle (the dose of 30 mg/kg/day was well below the level of 50 mg/kg/day which the FDA considered "safe"). I was not prepared for the severity of the reactions, and for obvious ethical reasons cannot perform any further human studies with aspartame. Over the ensuing years I have continued to see the multiple neurologic and psychiatric consequences of aspartame use. It can lower the seizure threshold and lead to an incorrect diagnosis of epilepsy, with subsequent inappropriate prescription of anticonvulsants. It can mimic or exacerbate symptoms of MS, it can paradoxically produce carbohydrate craving and weight gain. The worldwide epidemic of obesity and type 2 diabetes obviously has multiple causes, but I am convinced aspartame is a major factor.

The explosive increase in our knowledge base in the neurosciences I referred to earlier is a topic beyond the scope of this brief report, but to drastically oversimplify, we know that in a variety of psychiatric disorders there is a disturbance in the balance of certain neurotransmitters. Specifically, serotonin, norepinephrine, dopamine and acetylcholine are all major players.

Aspartame can affect the levels and balance of all these transmitters. It impairs the absorption of L-tryptophan, the major precursor in the synthesis of serotonin.

The phenylalanine from the dipeptide component of the aspartame molecule, is a major precursor in the norepinephrine-dopamine synthetic pathway. Recent research demonstrated that aspartame reduces acetylcholinesterase, an enzyme which breaks down acetylcholine—a key player in the central nervous system, with an important role in cognition and memory, and with a reciprocal, inhibitory relationship with dopamine.

We are not sophisticated enough at this point in time to fully understand all the implications of the neurochemical changes induced by aspartame, but as a busy clinician I see the profound impact on patients' lives on a daily basis. It can both produce and aggravate depression, in certain patients it can trigger manic episodes, it can produce or aggravate panic attacks. Some of my patients have experienced a complete cessation of panic attacks and needed no further treatment after they completely eliminated aspartame from their diet. Certain schizophrenic patients

> have experienced fewer auditory hallucinations or needed less antipsychotic medication after the elimination of aspartame. It is essential that there be much greater awareness of the hazards of this highly toxic substance!

Ralph G. Walton, MD
Medical Director, Safe Harbor Behavioral Health
Professor of Clinical Psychiatry, Northeastern Ohio University, College of Medicine
Adjunct Professor of Psychiatry, Lake Erie College of Osteopathic Medicine

Dr. Walton's aspartame study: "Adverse Reactions to Aspartame: Double-Blind Challenge in Patients from a Vulnerable Population"

http://www.mindfully.org/Health/Aspartame-Adverse-Reactions-1993.html

Dr. Walton's research on scientific peer-reviewed studies and funding: http://www.dorway.com/doctors.html#walton

Additional links regarding aspartame:
http://www.mpwhi.com/
http://www.mpwhi.com
www.mpwhi.com
www.dorway.com
http://www.wnho.net/
http://www.wnho.net
Aspartame Toxicity Center:

http://www.holisticmed.com/aspartame>www.holisticmed.com/aspartame

Betty Martini, DHum, Founder, Mission Possible International, 9270 River Club Parkway, Duluth, Georgia 30097, 770 242-2599

Note from Betty Martini:

This excellent new paper by Dr. Walton was distributed to an audience particularly concerned with psychiatric and behavioral problems. Also read Dr. Walton's comments about Abby Cormack of New Zealand who made world news when she was poisoned by aspartame in Wrigley's gum and about to be diagnosed as bipolar. Off aspartame, her symptoms disappeared. In New Zealand there were particularly sad aspartame/bipolar cases where families were wrecked. You can see Dr. Walton in the aspartame documentary "Sweet Misery: A Poisoned World," which is still being shown to audiences in NZ. It is alarming that in NZ they want Diet Coke sweetened with aspartame to be in "all" schools. The Minister of Health has been provided with "Report for Schools"

http://www.mpwhi.com/report_on_aspartame_and_children.htm
http://www.mpwhi.com/experts_on_aspartame_and_abby_cormack.htm

As to alternatives for schools, there is a new product just made available this month called Fiber 1, by Just Like Sugar, and it won't be in plastic. With aspartame (NutraSweet/Equal/Spoonful/Canderel/E951/Benevia, etc.) triggering psychiatric and behavioral disorders, it must be immediately removed from schools. Good nutrition is so important for children. Today they are medicated instead of educated

http://www.rense.com

Analysis Shows 100% of Independent Research Finds Problems with Aspartame

As far back as October 17, 1996, an analysis of peer-reviewed medical literature using MEDLINE and other databases was conducted by Ralph G. Walton, MD, chairman of the Center for Behavioral Medicine and professor of clinical psychiatry, Northeastern Ohio Universities College of Medicine. Dr. Walton analyzed 164 studies that were felt to have relevance to human safety questions. Of those studies, 74 studies had aspartame industry-related sponsorship and 90 were funded without any industry money.

Of the 90 non-industry-sponsored studies, 83 (92%) identified one or more problems with aspartame. Of the 7 studies which did not find a problem, 6 of those studies were conducted by the FDA. Given that a number of FDA officials went to work for the aspartame industry immediately following approval (including the former FDA commissioner), many consider these studies to be equivalent to industry-sponsored research.

Of the 74 aspartame industry-sponsored studies, all 74 (100%) claimed that no problems were found with aspartame. This is reminiscent of tobacco industry research where it is primarily the tobacco research which never finds problems with the product, but nearly all of the independent studies do find problems.

The 74 aspartame industry-sponsored studies are those which one invariably sees cited in PR/news reports and reported by organizations funded by Monsanto/Benevia/NutraSweet (e.g., IFIC, ADA). These studies have severe design deficiencies which help to guarantee the "desired" outcomes. These design deficiencies may not be apparent to the inexperienced scientist. Healthcare practitioners and scientists should print out the all of the documents on the Monsanto/NutraSweet Scientific Abuse web page, the Scientific FAQs web page, and the Aspartame Toxicity Reaction Report Samples. Please refer scientific questions to mgold@holisticmed.com.

Aspartame (NutraSweet) Toxicity Information Center Main Page.

Reprinted with permission from Operation Mission Impossible

THE BITTER TRUTH ABOUT ASPARTAME AND NEOTAME
Illegal Actions Gained FDA Approval

FDA approved the new genetically engineered molecule in 1974, following a spate of criminal acts committed by G.D. Searle pharmaceutical employees and contractors. That approval was rescinded before aspartame got to market, because it was revealed a component of aspartame [diketopiperazine] caused brain tumors. Tests were "falsified." Second approval occurred in 1981 under President Reagan's new FDA Director, Dr. Arthur H. Hayes. In 1983, Hayes' office approved the new molecule for aqueous solution [soft drinks, etc.] and three months later, Dr. Hayes left government, went to work for the NutraSweet public relations firm, Burson Marsteller for $1,000 a day. Hayes continues to craftily dodge all media requests for interviews on the topic. In fact, many government officials connected with this issue have defected—going to work for the industry whose products they were entrusted to approve or reject.

In the laboratory, aspartame produced:

* Brain Tumors
* Breast Tumors
* Uterine Tumors
* Pancreatic Tumors
* Seizures
* Deaths

These went unreported, as did the deaths and seizures of other animals in the original tests. Tumors were removed from animals put back in test as tumor free, animals who died were brought back to life on paper when results officially submitted by G.D. Searle pharmaceutical to FDA.

Aspartame changes DNA:

In tests, the third generation of pups born to animals fed aspartic acid [component of aspartame] were born:

1. Morbidly obese and
2. Sexually Dysfunctional
3. Components of aspartame and its breakdown products can adversely affect the brains and central nervous systems of children and adults who consume it.
4. "When you harm the brain, you harm the very expression of one's self."—Peter Breggin, MD [Psychiatrist - Bethesda, MD]

5. **Reported psychological symptoms:**
6. * Suicidal Depressions
7. * Panic Attacks and Anxiety [PAD] - Manias
8. * Sleep Disorders
9. * Severe Mood Disorders [rages, mood swings]
10. * Brain Chemical Imbalance
11. * Brain Wave Malfunctions [shows up in EEGs]
12. * Personality Disorders
13. * Hallucinations
14. * Aberrant Behaviors

Medicine in Our Food?

Aspartame Components and Breakdown Products:

Phenylalanine 50%—Lowers the seizure threshold. Causes mental retardation in some. Blocks production of serotonin [key neurotransmitter which controls: Moods / Sleep Patterns / Satiety] Cultured in e-coli bacteria in the lab.

Aspartic Acid 40%—Caused lesions or holes in the brains of lab animals. Neuroexcitatory [excites brain cells to death] amino acid. Causes motor-neuronal disorders in studies.

Methanol 10%—Also known as "methyl alcohol," "moonshine," "Sterno." Damages the liver and eye. Two teaspoons can be lethal to humans. [Not processed the same in humans and animals, so lab tests do not show full impact of toxicity] Breaks down into formaldehyde [embalming fluid] and formic acid [venom in insect stings] at temperatures exceeding 86 degrees Fahrenheit. [Body temperature is 98.6 F] Implicated in birth defects and fetal alcohol syndrome in newborn infants. As a constituent of other foods in nature, it is found in combination with ethyl alcohol, which counteracts or neutralizes the toxic effects of methanol as it is metabolized by the body. There is no ethyl alcohol in aspartame, therefore methanol in aspartame is in "free form" and is immediately absorbed into the bloodstream. For every molecule of aspartame, there is a molecule of methanol released. Classic signs of methanol poisoning include:

lethargy, confusion, leg cramps, back pain, severe headache, abdominal pain, slurred speech, fainting, visual loss/blindness, labored breathing.

Diketopiperazine [DKP]—Caused brain tumors in laboratory tests. Thirteen out of 320 lab animals developed brain tumors in testing. Aspartame breakdown products cross the blood-brain barrier to damage the brain.

Aspartame is known to exacerbate or trigger onset of the following medical conditions:

* Epilepsy
* Parkinson's
* Alzheimer's
* Multiple Sclerosis
* Chronic Fatigue Syn.
* Lymphoma
* Fibromyalgia / Eosinophilia Myalgia
* Mental Retardation / Birth Defects
* Diabetes / Hypoglycemia
* Graves's Disease
* Heart Disease
* Lung Disease
* Liver Disease
* Kidney Disease
* Brain Tumors [astrocytoma/glioblastoma]
* Pancreatic Disease
* Kidney / Adrenal Disease
* Arthritis
* Blindness
* Tinnitus
* PMS
* Carpal Tunnel
* Lyme Disease
* Muniere's Disease
* Other: Rare / Hard to diagnose disorders

[Aspartame has been called a "systemic" toxin—which means it may virtually adversely affect the function of every organ of the body. The effects are "cumulative" and do not show up in short-term testing.]

Aspartame and the Media:

1990, The Guardian [London newspaper] was sued for publishing headline story—"NutraSweet Faked Tests to Get Sweetener Approved." Later settled out of court. NutraSweet sent intimidating press releases to all US media warning they were suing.

Documented instances are on record where the manufacturers of aspartame threatened to sue and/or withdraw advertising dollars, from media. Health care professionals, scientists and medical schools have been "bribed" or intimidated into silence.

Aspartame and Flying Safety:

Many military, general aviation and commercial airline pilots have lost medical certification to fly based on seizures which occurred while they were ingesting aspartame. Grand mal seizures have been reported in flight in the cockpits of commercial airliners. A pilots' hotline in Dallas, TX was established by ACSN, in 1988 for the anonymous reporting of adverse reactions and safety-of-flight incidents. USAF Flying Safety magazine published warnings re: aspartame use by pilots. Pilot's publications around the globe have warned readers. The FAA will not send out an official memo—because the FDA refuses to recall aspartame as a safety hazard to consumers. Is aspartame the unacknowledged "terrorist" in every cockpit on every flight?

Legal:

ACSN cofounders Turner and Stoddard have both qualified in court as medical Expert Witnesses. Many consumer lawsuits have been dropped or settled out of court since the mid-eighties. The Washington D.C. Supreme Court refused to hear a case brought against the FDA by James Turner, Esq. Aspartame is illegal because it violated The Delaney Clause, which states no substance can be approved that is shown to cause cancer.

In 1995, a stealth law crept across the land, which made it illegal to say anything disparaging about a perishable food product—example: yogurt sweetened with aspartame. Oprah was later sued under this law. [Agricultural Defamation Act]

In 1985, G.D. Searle and NutraSweet Co. became wholly-owned subsidiaries of Monsanto Chemical in St. Louis. Controversial Supreme Court Judge, Clarence Thomas is a former Monsanto attorney. Monsanto sold their sweetener divisions to current owner, the J.W. Childs Partnership. 1998, Monsanto applied for FDA approval for a monster molecule, "based on the aspartame formula" with one addition: 3-dimethylbutyl [listed on EPA's most hazardous chemical list]. Thus, Neotame becomes 13,000 times sweeter than sugar.

July 5, 2002—Neotame, Monsanto's super bio-manipulated molecule [newest fake sweetener] was approved by FDA over formally registered objections of the Aspartame Consumer Safety Network and others. Long-term effects on humans are unknown. Aspartame is in over-the-counter drugs like: Tums, Pepsid AC, Metamucil, Alka Seltzer Plus, tooth whiteners, breath mints/strips and more.

Parents—the FDA does not protect your children.

Greed and avarice have made government officials, industry and mainstream science, turn a deaf ear to the truth. Children are at least 4 times more susceptible to aspartame toxicity because of the developing central nervous system. Read labels. If it says: "phenylalanine" dont buy it. Insist your child's pediatric medical records contain this warning—"Never prescribe any medications containing phenylalanine [always present in aspartame and neotame]." Aspartame is a multi-billion dollar-a-year industry in more than 7,000 food products, chewing gums, children's antibiotics and meds, children's vitamins and pain medications for kids. In the lab, aspartame caused tragic birth defects.

Action Agenda:

* Get off the substance for a period of 4-6 weeks to see your physical and mental condition improve.
* If you have sought medical treatment for any of these symptoms and you are/were an aspartame user—have your doctor enter into your official medical record that you believe aspartame may have caused some or all of your symptoms and that you are ceasing use of the sweetener and want to be "monitored" by your health care team as you get better. Very important.
* Join over 10,000 concerned consumer members of Aspartame Consumer Safety Network worldwide who are fighting this battle to educate the consuming public about artificial sweeteners that masquerade as "natural."
* Take all aspartame products back to the store and exchange them for healthy "real" food. This sends a message back to the manufacturer that we will no longer tolerate neurotoxins in our foods and children's items. Register a powerful complaint where it counts.
* Write letters to the FDA and food manufacturers who use aspartame. Copy the editor of your local newspaper, stating your displeasure about the conspiracy of silence and illicit cover up of the facts by government and industry. The "real villains" in today's society are those who poison our food supply and that of our children—for profit.

Published by Aspartame Consumer Safety Network
Mary Nash Stoddard, Founder and President
P.O. Box 780634, Dallas, TX 75378-0634
214.352.4268
Website: http://web2.airmail.net/marystod/ email: marystod@airmail.net

Additional references on aspartame

http://www.holisticmed.com/aspartame/aspfaq.html
http://www.holisticmed.com/aspartame/adverse.txt
http://www.holisticmed.com/sweet/http://www.newviewtv.com/channel6.ram
Part 1 http://www.newviewtv.com/sweetp2nol.ram
Part 2 http://www.holisticmed.com/aspartame/fm.html
http://www.dorway.com/peerrev.html
http://www.sweeypoison.com
http://www.dorway.com/bressler.txt
http://www.sunsentpress.com
http://www.dorway.com/betty/brainc.txt
http://www.wnho.net/pseudotumor_cerebri.htm
http://www.mpwhi.com/complete_bressler_report.pdf
http://www.dorway.com/raoreport.txt
http://www.soundandfury.tv/pages/rumsfeld2.html
http://www.holistic.com/aspartame/abuse/
http://www.rense.com/general/asp.htm

Chapter 8

Gulf War Syndrome

Many veterans who served quite a few tours of duty, either in Afghanistan or Iraq, are complaining of a multitude of health related symptoms. It is a known fact that tons of diet drinks were sent to our troops serving in the Gulf. The problem arises when these drinks are stored in sometimes temperatures that can be as high as 120 degrees Fahrenheit in the desert. Since aspartame breaks down into its components of phenylalanine, aspartic acid, methanol, and trace amounts of DKP (diketopiperazine) at temperatures of only 86 degrees, the drinks already were composed of these compounds. As was noted earlier, these compounds can become toxic, depending on the amount consumed. Being in a high-temperature, dry-climate region, large amounts of fluids must have been consumed.

About seven hundred thousand troops served in the Gulf wars, and about ninety thousand have desert storm syndrome. The reported symptoms are memory loss, respiratory and heart problems, chronic fatigue syndrome, joint aches, confusion, depression, muscle pain, headaches, equilibrium problems, insomnia, vision loss, and birth defects.

This is very interesting since all of these reported symptoms were pretty well mentioned by the Department of Health and Human Services when they tabulated a list of reported symptoms to the FDA. Another study on aspartame toxicity effect reported similar symptoms. (See provided lists previously.) FDA has since shut off complaints regarding aspartame complaints.

Dr. James Bowen explained the toxic effect in the following way: "Aspartic acid, the excitotoxin compound of aspartame does not cross the blood-brain barrier, but is secreted into the cerebral spinal fluid by the choroid plexus. [This is an area on the ventricles of the brain where cerebrospinal fluid is produced. Liquid filters through these cells from blood to become cerebrospinal fluid. Ventricle is a cavity in the brain that is an enlargement of the central canal of the spinal cord containing cerebrospinal fluid.] There in the brain's lower area and upper terminus of the spinal is where Lou Gehrig's disease, Parkinson's disease and multiple sclerosis damage is most prominent. These critical locations are bathed in the toxin as it is removed from the blood. From the third to fourth ventricles there is a narrow canal, called sylvian aqueduct, which fills with this secretion and washes the roof of the hypothalamus."

Two independent studies appeared in the *Neurology Journal* (Sept. 23, 2003, issue) both reported an above average occurrence of amyotrophic lateral sclerosis among Persian Gulf War deployed veterans. The deployed troops developed Lou Gehrig's disease more frequently and at an earlier age. Aspartame does not know age or geography. Deployed troops in the Persian Gulf were consuming lots of diet pop that sat for long periods of time at about 120 degrees in the Arabian sun, according to reports by some troops.

"Time Life published an article on bipolar disease, well known to be triggered by aspartame because of the depletion of serotonin, which causes manic depression, panic attacks, rage, and violence. This is an adult disease, but the article interestingly point was that children were developing bipolar disease. 40% of children in US are using it. Movie star Michael Fox wanted to know why he was developing Parkinson's disease, an old man's disease at age 30. He was a Diet Pepsi spokesman and addicted to it." (*http://www.medical-library.net/aspartame_disease.html*)

The Alzheimer's Association says, "50% of all nursing home beds have Alzheimer patients, including mature baby boomers. A hospice nurse reported they were getting 30 year olds with Alzheimer's, and aspartame is escalating this disease" (book by Dr. H. Roberts, *Defense Against Alzheimer's Disease*, 1-800-814-9800)

"So what triggers Lou Gehrig's disease (ALS)? The answer ASPARTAME!" Dr. Brown wrote FDA nineteen years ago. "Aspartic acid is a neuroexcitotoxin present in damaging amounts in soft drinks. Simple logic tells one that it will vastly increase the metabolism of methyl alcohol to formaldehyde. This corresponds well with the symptomalogics often experienced such as ALS (Lou Gehrig's Disease), bulbar palsies, neurohormonal disorders . . . Diketopiperazine issue remains totally unresolved and dangerous . . . An Air Force pilot traced his pattern of tremors and seizures for two years to consumption of NutraSweet. When he traveled in areas where diet drinks were not available, he was free of the symptoms. When he resumes diet drinks, symptoms returned. In January 1990, Plane and Pilot Magazine featured an article on drugs and alcohol, and also discussed food additives. It mentioned NutraSweet and Equal and that aspartame contains 10% methanol which is a poisonous substance. The article disclosed that methanol destroys the brain, a little bit at a time, and that its effect is cumulative. Depending on the individual, the reaction can be immediate or later on in life."

For further information, you can contact the Aspartame Consumer Safety Network in Dallas, Texas. Contact Mary Nash Stoddard at 1-800-352-4268

Noted neurosurgeon Dr. Russell Blaylock, MD, wrote in his book *Neuroexcitotoxins: The Taste That Kills*, about diseases triggered by such excitotoxins as MSG (monosodium glutamate) and aspartic acid. He quotes Dr. John Olney pointing out the irony of a food industry practice.

Consuming a lot of aspartame may inhibit the ability of enzymes in your brain to function normally according to a new review by Scientists from the University of Pretoria and the University of Limpopo.

Here is what the report by these institutions reported:

> Consuming a lot of aspartame may inhibit the ability of enzymes in your brain to function nor-

mally." The review found also that high doses of the sweetener may lead to neurodegeneration. It has also been found that aspartame consumption can cause neurological and behavioral disturbances in sensitive individuals. The review found a number of direct and indirect changes that occur in one's brain as a result of high consumption levels of aspartame, including disturbing effects on: 1) metabolism of amino acids, 2) protein structure and metabolism, 3) integrity of nucleic acids, 4) neuronal function, 5) endocrine balances.

The breakdown of aspartame causes nerves to fire excessively, which can indirectly lead to high rate of neuron depolarization. Strangely enough, this is what other independent studies have also indicated. Despite all these growing concerns, neither the European Food Safety Authority (EFSA) or the US Food and Drug Administration (FDA) have changed their guidelines regarding the safety of the ingredient or intake advice. There is a plethora of evidence regarding the lethal health effects of this ingredient, approved under false pretenses by the FDA. Is there a much more sinister agenda by these agencies? You be the judge, dear reader.

(Reprinted with permission from Operation Mission Impossible)

Gulf War Veteran Resource Pages

Editor's Note: Aspartame Consumer Safety Network, Worldwide Pilot Hotline, and Operation Mission Impossible are three worldwide organizations warning the world about aspartame poisoning.

Of the 697,000 who served in the Gulf War, 43,000 have desert storm syndrome. Reported ailments included the following:

Memory loss	Muscle pain
Respiratory and heart problems	Headaches
Chronic fatigue syndrome	Equilibrium problems
Joint aches	Insomnia
Confusion	Vision loss
Depression	Birth defects

In September 1995, *Sixty Minutes* did an expose on the syndrome and identified another symptom of the "burning tongue"! The Chronic Fatigue and Immunologic Disease Syndrome Network says six thousand perished!

A soldier now retired in Huntsville, Alabama, called with his story. He said he suffered terribly with loss of memory, headaches, and chronic fatigue. I asked if he used NutraSweet and he said, "The soda pop companies sent free diet drinks to the Gulf because of a scarcity of water, so we drank diet drinks all day long."

He said the diet drinks sat on pallets for up to eight weeks in 120-degree Arabian sun. I said, "Think of which ones drank the diet soda and which ones drank water, and tell me which ones have the syndrome."

He replied with little hesitation: "As I recall, now that you mention it, the ones who are sick are the ones I remember drinking the diet soda."

Aspartame, sold as NutraSweet/Equal/Spoonful/etc., is a chemical poison. It is a molecule composed of three components: aspartic acid, phenylalanine, and a methyl ester that converts to methanol, or wood alcohol, a severe metabolic poison. In your body, the wood alcohol converts to formaldehyde and then formic acid (ant sting poison). The phenylalanine breaks down into DKP, a brain tumor agent. There are ninety-two documented symptoms but here are some of the most common:

Symptom	**Causes:**
Memory loss and confusion	Neurotoxic amino acids passing blood-brain barrier and deteriorating neurons of the brain
Chronic fatigue syndrome	Methanol breaks down immune system
Joint pain	Methanol and other components
Equilibrium problems	Methanol toxicity
Heart and respiratory symptoms (tachycardia, shortness of breath, unexplained hypertension)	Aspartame affects certain neurotransmitters that influence the brain's respiratory center. Methanol also affects the heart. Problems caused by multiple components.
Anxiety attacks, manic depression, suicidal tendencies, violence, etc.	The phenylalanine in aspartame breaks down the seizure threshold of the brain and blocks serotonin production. Researchers attribute low serotonin to manic depression, panic attacks, etc.
Insomnia	Phenylalanine (also causes hallucinations)
Vertigo, numbness, tinnitus	Methanol toxicity (mimics MS)

Birth defects (Dr. Louis Elsas, Emory University professor of pediatrics) at a hearing before US Senate Committee on Labor and Human Resources, examining NutraSweet and safety concerns 11/3/87): "I have spent 25 years in biomedical sciences, trying to prevent mental retardation and birth defects caused by excessive phenylalanine. And therein lies my basic concern that aspartame is in fact a well-known neurotoxin and teratogen causes birth defects) which in some as yet unidentified dose will both reversibly in adult and irreversibly in developing child or fetal brain produce adverse effects."

Dr. H. J. Roberts says he is concerned about birth defects "because of histories of reports given me about severe problems in the fetus or infant of parents (including fathers) who consumed much aspartame at the time of conception and/or during pregnancy. Several animal studies support such concern."

A lady on Internet complained about a "burning sensation in her mouth." She called the Pepsi Co. and they said it was due to the sweetness breaking down. They sent her a free case. It is methanol that gives the drink sweetness, and when it breaks down to formaldehyde, it burns your tongue and is bitter.

At eighty-six degrees, aspartame liberates methanol in the can. The soldiers drank a toxic cocktail!

The common symptoms of aspartame toxicity are headaches, memory loss, confusion, and disequilibrium. These are principle symptoms of desert storm syndrome. To our great good fortune a world-famous physician and researcher has been investigating this syndrome for more than a decade and published two books and many clinical reports on the subject. He is H. J. Roberts, MD, FACP, FCCP, once distinguished as the most outstanding doctor in the United States. Roberts reports that when aspartame was approved his patients, particularly diabetics, increasingly presented with memory loss and confusion. He explains that the amino acids in aspartame, aspartic acid, and phenylalanine are neurotoxic as isolates, unaccompanied by other amino acids normally present in our foods. As isolates, they penetrate the blood-brain barrier, attack and destroy the neurons (brain cells), and produce multifold neuropsychiatric complications including depression, brain tumors, Alzheimer's, seizures and a long list of other disabilities. Dr. Roberts has named this collection of pathogenic responses *aspartame disease*. Considering the thousands of reports of these identical symptoms from veterans of the Gulf War the question is, is desert storm syndrom simply another name for aspartame disease?

To exacerbate this chemotoxic assault on the wellbeing of the desert storm troops, they received multiple vaccine injections, which consistently invoke immunological reactions in a percentage of the injectees. Such a deadly bouquet when combined with a daily liter of NutraPoisoned diet drinks!

Betty Martini, Operation Mission Possible

Mission Possible is a volunteer force in 50 states distributing a warning flyer worldwide on aspartame. We are dedicated to the proposition that we won't be satisfied until death and disability are no longer considered an acceptable cost of business.

We welcome more volunteers to distribute this worldwide alert and pass the torch of knowledge. We look forward to the day we will be able to hold up this torch to the world and say MISSION ACCOMPLISHED!

Those wanting more information on Aspartame can email me for instructions to the auto-responder where you can access our warning flyer and other information to print out, plus other posts on other subjects.

Betty Martini - betty@pd.org

Chapter 9

Autism

Autism according to some definitions is a developmental disorder that appears by age three and it is variable in expression. According to the definition from *Encyclopedia Britannica* autism is a "neurological disorder that affects physical, social, and language skills." First described by Leo Kanner and Hans Asperger in the 1940s, the syndrome usually appears before two years of age. Autistic infants appear indifferent or averse to affection and physical contact. They may be slow in learning to speak and suffer episodes of rage or panic; they may also appear deaf and display an almost hypnotized fascination with certain objects. Autism is often characterized by rhythmic body movements, such as rocking or hand-clapping and by obsessive desire to prevent change in daily routines. Autistic individuals may be hypersensitive to some stimuli (e.g., high-pitched sounds) and abnormally slow to react to others (e.g., physical pain). The disorder is three to four times more common in males. Though postnatal factors such as lack of parental attention were once blamed, it is now known that autism is the result of abnormalities in the brain structure."

Autism is often referred to as an autism spectrum disorder (ASD). People with autism represent a broad group of individuals with varying degrees of disability. These individuals have problems that are classified in three different areas. These are social interaction, communicating with others, and behavioral problems. Each case is different. The complex neurobehavioral disorder covers a wide range of levels of impairment, skills, and mental handicap. The severity from such handicap limits an individual from a normal life to a devastating disability requiring potentially an institutional care.

Autism was first discovered in the early 1940s by two doctors, child psychologists, working independently. The word *autism* comes from the Greek word *self*, and it was used to describe children who seemed enclosed in their own solitary worlds. Initially the occurrence of autism was about 1 in 10,000 in 1940, but in the last few decades, autism has become increasingly more commonly diagnosed childhood developmental disorder. In 2009, Centers for Disease Control Prevention reported that 1 in 110 children were affected by autism. Statistics from the US Department of Education and other government agencies indicate that autism diagnoses are increasing at the rate of 10 to 17% per year. In 2010, the number of autistic cases was at 1 in 88 children. For 2014 it is estimated that 1 in 40 cases for boys, and 1 in 60 cases for girls. Autism is not limited to any ethnic, racial, or social background. Boys are four times more likely to have autism than girls. The researchers have a difficult time in determining the cause of autism. The question they are asking is if the increase in cases due to better diagnostic technique or whether it is a true increase due to environmental factors.

The most recent version of the *Diagnostic and Statistical Manual of Mental Disorders*, Fifth Edition (*DSM-5*) has just a single category for the diagnosis of an autistic disorder—autism spectrum disorders. These include the following:

* Autistic disorder: It is what people think of when they hear the word "autism." It refers to problems with social interactions, communications, and imaginative play in children younger than three years.
* Asperger's syndrome: It was named after one of the original doctors that uncovered this disease. These children don't have a problem with language. They tend to score in the average or above-average range on intelligence tests but have the same social problems and limited scope of interests as children with autistic disorder.

* Pervasive developmental disorder (PDD): It is also known as atypical autism. This is a kind of catch-all category for children who have some autistic behaviors but who don't fit into other categories.
* Childhood disintegrative disorder: These children develop normally for at least two years and then lose some or most of their communication and social skills. This is an extremely rare disorder and its existence as a separate condition is a matter of debate among many mental health professionals.
* Rhett syndrome: Children with Rhett syndrome, primarily girls, start developing normally but then begin losing their communication and social skills. Beginning at the age of 1 to 4 years, repetitive hand movements replace purposeful use of hands. Children with Rhett syndrome are usually severely cognitive impaired. Since that classification in autism spectrum disorder, a meeting of the American Psychiatric Association in San Francisco in May 2013 eliminated the Asperger disorder as a separate condition.

Autism spectrum disorder (ASD) is considered a neurological condition that affects the way the brain functions, resulting in difficulties with communication and social interaction and unusual pattern of behavior, activities, and interests. Some people can function well, while others are locked into their own separate world.

Symptoms of Autism

There is actually no single symptom that can lead to a diagnosis of autism. Anyone showing a number of the following characteristics and behaviors would likely be diagnosed as being with ASD. These can be the following:

* Shows no interest in other people
* May be interested in people but does not know how to talk, interact with, or relate to them

* Has difficulty initiating and maintaining a conversation
* Is slow developing speech and language skills, which may begin to develop and then be lost or may never develop fully
* Has difficulty interpreting nonverbal communication, such as social distance cues or the use of gestures and facial cues, like smiles, that most of us take for granted
* Repeats ritualistic actions such as spinning, rocking, staring, finger flipping, and hitting himself
* Has restricted interests and seemingly odd habits, like focusing obsessively on only one thing, idea, or activity

People with ASD may also have these secondary problems such as the following:

* Neurological disorders, including epilepsy
* Gastro-intestinal problems—due to inflammation
* Fine and gross motor deficits
* Anxiety and depression

Interesting to note that a lot of these symptoms, conditions are also attributed to fluoride, aspartame, monosodium glutamate (MSG), and artificial colors and tastes. How about the GMO foods? Did any of the so-called researchers ever consider asking the parents as to consumption of products containing above poisons during pregnancy? Infant formulas? What about all the numerous vaccinations that contain poisons such as ethyl-mercury (thimerosal) and aluminum, both classified by numerous doctors and scientific studies as neurotoxins. The number of infant vaccinations has risen to almost 34. Can such early developing brain handle all these toxins? What also has become apparent that some whistleblowers working for the pharmaceutical companies, have come forward stating that they were told to lie regarding vaccines that cause autism. More on that subject under the topic of vaccination.

During the last meeting of Autism Speaks Association, ten toxins were identified in the environment that can contribute to the autism spectrum. Interesting to note that fluoride, aspartame, and now neotame, MSG, were omitted from the list. The toxins listed are (1) lead, (2) methylmercury (strange since ethylmercury is in vaccines and toxicity behaves in similar function as methylmercury), (3) PCBs, (4) organophosphate pesticides, (5) organochlorine pesticides, (6) phthalates, (7) bisphenol A, (8) automotive exhaust, (9) polycyclic aromatic hydrocarbons, and (10) brominated flame retardants.

Nothing was mentioned about fluoride and its many applications as fluoride in water, organic compounds in pesticides, antibiotics, and agricultural fumigants. In the section for fluoride discussion, Chinese studies clearly indicate brain cell damage among one of the adverse effects. The same can be said about MSG. It is covered in a separate section of this book. Extensive discussion and numerous links provided for aspartame, regarding adverse physical and mental health effects. Ever since all these toxic substances have been incorporated into our life, autism has dramatically increased. Yet our so-called health organizations like EPA, AMA, FDA, USDA, NIH, and NCR still close their eyes to scientific studies that indicate the toxic effect of these poisonous substances.

http://www.autismspeaks.org/what-autism
http://www.autism-society.org/what-is/
http://www.webmd.com/brain/autism/
http://www.webmd.com/brain/autism/autism-symptoms

Autism toxins

"We all have hundreds of chemicals in our bodies today that didn't exist a few decades ago," says Miller. "And we're seeing increases in learning and developmental disabilities as well as many other chronic diseases. Currently, one in six children under the age of 18 has some kind of learning, or developmental, or behavioral disorder."

While there's debate about just how much of that is an actual increase, and how much may be due to factors like better diagnosis, people who have worked with children for a long time are seeing a change, according to Miller. "I talk to a lot of teachers," she says, "and any of them who have been in the classroom the last 20 or

25 years will tell you, 'I used to have one kid or two kids who had learning problems or were disruptive, and now, half my class has behavioral issues.' That's not necessarily all because of environmental exposures, but genes don't change that quickly. So social, nutritional, and environmental factors have got to be playing a significant role."

The medical profession still does not want to admit as to the cause of autism for whatever reason. The predominant indicating factors point squarely at all the environmental factors like fluoride, aspartame, MSG, the numerous pesticides, etc. How about comparing the very young children to children to Amish families? Did anyone ever make comparison like that? I suppose the resultant results would negate all the propaganda regarding fluoride, and all the other toxins fed to general population in food, water, and air.

More vulnerable, pound for pound

While much research has been done, one point has become quite clear: children are far more vulnerable to toxins in the environment than adults. A new study done by researchers at the University of Chicago revealed that autism and intellectual ability (IQ) rates are linked to harmful environmental exposure factors during congenital development. This study confirmed what was already reported in mid nineteen nineties by the Chinese studies and then verified by the Harvard study. According to Professor Andrey Rzhetsky of genetic medicine and human genetics at the University of Chicago: "Essentially what happens is during pregnancy . . . there are certain sensitive periods where the fetus is very vulnerable to a range of small molecules—from things like plasticizers, prescription drugs, environmental pesticides and other things. Some of these small molecules essentially alter normal development. Autism appears to be strongly correlated with rate of congenital malformations of the genitals in males across the country, this gives an indicator of environmental load and the effect is surprisingly strong. The strongest predictors for autism were associated with the environment; congenital malfunctions on the reproductive system in males."

One of the more interesting facts about autism is that it has remained fairly constant in Europe, Japan, and Australia. This coincides with significant restrictions or complete bans of fluoride, sale of GMOs, and the pesticide that go with them. In the US, EPA just raised the allowable limits on glyphosate, which WHO (World Health Organization) just declared a potential carcinogen. A study published in the *Journal Reproductive Toxicology* successfully identified the presence of pesticides—associated with genetically modified foods in maternal, fetal, and nonpregnant women's blood. They also found presence of Monsanto's Bt toxin. Reading multiple studies, they come to the conclusion to do further studies associated with GMOs and the associated pesticides. How about doing the studies *prior* to release for general consumption, not *after*? The Standing Committee of European Doctors stated, "Chemical pollution represents a serious threat to children, and to Man's survival." A study by US Geological Survey done on Mississippi air and rain between 1997 and 2007 revealed "that Roundup herbicide and its toxic byproduct AMPA were found in over 75% of the air and rain samples tested in 2007."

A study published in peer-reviewed journal *Translational Neurodegeneration* revealed evidence supporting an association between increasing organic-mercury exposure from thimerosal contained in childhood vaccinations to risk of ASD. Bottom line is that vaccines are full of toxins, administered at critical stages of development, including during pregnancy. Linking autism to toxin exposure further signifies in evaluating all the multiple vaccines that babies are bombarded with at birth and the next eighteen months of their lives.

What is really alarming is that the vaccine manufactures and health officials/authorities have known about and covered up the hidden dangers associated with vaccinations in order to protect herd immunity. Documents obtained by Lucija Tomljenovic, PhD, from the Neural Dynamics Research Group at the University of British Columbia, reveal that vaccine manufacturers, pharmaceutical companies, and health authorities have known about the multiple dangers associated with vaccines. They chose to withhold this critical information from the parents and the public in general.

Children are also more vulnerable because they are still growing. Key organ systems such as the brain and nervous system, lungs, and reproductive organs are all still developing rapidly in the first few years of life, making them susceptible to interference from toxic chemicals. In addition, the kidneys and liver are not fully developed and can't detoxify harmful substances as well as those of adults. Children are exposed to environmental toxins in various ways. School buses, for example, which shuttle millions of children between home and school every day, routinely trap alarmingly high levels of diesel exhaust inside, according to a study conducted by the National Resources Defense Council, the Coalition for Clean Air, and the University of California, Berkeley. And numerous studies have shown that diesel fumes cause cancer, particularly lung cancer. In one study of baby foods sold in the United States, more than one-half of all samples contained detectable levels of pesticides. Nearly one fifth of baby food jars examined contained two or more pesticides. Food is not the only item children put in their mouths. Children, by nature, explore their environment, often by putting dirt, paint, or other nonfood substances into their mouths, potentially exposing them to environmental toxins. Exposure can come from unexpected—and sometimes tragic—sources as well.

http://www.nejm.org/doi/full/10.1056/NEJMoa1307491

http://www.mdpi.com/1099-4300/15/4/1416,

http://articles.latimes.com/2007/may/25/nation/na-fetuses25,

http://www.truth-out.org/news/item/23267-autism-nation-americas-chemical-brain-drain

Reprinted with permission from Natural News:

Moms of autistic, vaccine-damaged children mount wave of online protest against CDC research fraud

Friday, September 05, 2014 by: Jonathan Benson, staff writer

Tags: autistic children, vaccines, CDC

Chemistry professor corroborates whistleblower's claims; links mercury in vaccines to autism,

(NaturalNews) A groundswell of backlash against government corruption has been unleashed across the web following the revelation that the Centers for Disease Control and Prevention (CDC) lied about data linking vaccines and vaccine ingredients to autism. The hashtags "#CDCwhistleblower" and "#hearthiswell," a reference to a global campaign aimed at "breaking the science on vaccine violence," are bursting forth across Twitter, Facebook and YouTube, drawing attention to the US government's evil vaccine war against the most vulnerable members of society, our children.

Moms everywhere are posting videos on YouTube bearing these hashtags, in which they talk about their own children who were damaged by vaccines, warning others to be cautious before injecting their babies without researching the matter on their own. One mother of eight children, for instance, talks about how, among the six that were vaccinated, three have autism, and the others have a variety of mood and behavioral disorders, as well as gastrointestinal and other physical health problems.

"Out of my vaccinated children, three have autism," she states. "One has ADHD [attention-deficit hyperactivity disorder], one has a severe language disorder, and one has severe mood

> swings. They have also suffered from asthma, eczema, chronic ear infections, gastrointestinal disorders, urinary tract infections, psoriasis, food allergies, chemical sensitivities."
>
> "My two unvaccinated children have had none of their siblings' disorders—their health has been excellent," she adds.

CDC whistleblower tells all, admitting his agency lied about the fact that vaccines cause autism

These and many other similar videos began pouring in after Dr. William Thompson, a former high-level scientist at the CDC, broke the silence on fraud at the CDC that led to critical data linking vaccines to autism being censored and withheld. Dr. Thompson helped published a fraudulent study in the journal *Pediatrics* back in 2004 that the government and mainstream media claim is "evidence" that vaccines are safe.

But Dr. Thompson recently came clean to Dr. Brian Hooker from the Focus Autism Foundation, who also has vaccine-injured children. The two joined forces with leading gastroenterologist Dr. Andrew Wakefield to produce a short video explaining how the CDC altered study data to make it seem as though vaccines and autism are unrelated, when in fact they are deeply connected.

Don't trust the CDC, do your own research, and skip the vaccines

No longer are parents of vaccine-injured children going to be chided by the pseudoscientific elite who claim that the debate is over, and that vaccines are safe. As you will see in the aforementioned video, this particular mother purchased several books about vaccines, presumably after the fact, that gave her an honest insight into their true dangers—something that few mainstream doctors will ever recommend.

This is not simply blind opposition to vaccines "because of Jenny McCarthy," or other such straw man accusations that often get flung at parents who resist the religion of vaccines. More parents these days are doing their homework and coming to the sound conclusion that vaccines are dangerous, something that has now been solidified by the CDC whistleblower and that will continue to spread thanks to efforts like this hashtag campaign.

To watch other videos of parents with vaccine-injured children, check out the "Hear This Well" YouTube page here: *YouTube.com*.

Sources for this article include:

https://www.youtube.com
http://www.youtube.com
http://www.vimeo.com
http://science.naturalnews.com
http://www.naturalnews.com/046750_autistic_children_vaccines_CDC.html#ixzz3CU78daax

ADHD

What is ADHD?

ADHD is Attention Deficit Hyperactivity Disorder. According to David Rabiner, PhD, at the University of Duke. "ADHD is a disorder characterized by persistent pattern of inattention and/or hyperactivity/impulsivity that occurs in academic, occupational, or social settings. Problems with attention include making careless mistakes, failing to complete tasks, problems staying organized and keeping track of things, and becoming easily distracted. Problems with hyperactivity can include excessive fidgeting and squirminess, running or climbing when it is not appropriate, excessive talking, and being constantly on the go. Impulsiveness can show up as impatience, difficulty awaiting one's turn, blurting out answers, and frequent interrupting. Although many individuals with ADHD display both inattentive and hyperactive/impulsive symptoms, some individuals show symptoms from one group but not the other." At best it is a subjective observation requiring extensive training to recognize. Many individuals can display these behaviors, especially children, but it is quite difficult to differentiate

ADHD behavior from normal type behavior. There is a marked difference between an energetic child and one who causes excessive talking and has difficulty completing homework or other assignments. For a child with ADHD, these problems cause significant impairment in daily functions. It manifests itself more prominently in a classroom. Sometimes the child can do well in school and home and sometimes not at all. The performance can fluctuate. This does not mean that any child who shows fluctuations in school performance has ADHD. There are a variety of reasons for erratic performance. To the primary above symptoms, children with ADHD can display a number of other difficulties. It is important to note that these associated difficulties are not symptoms of ADHD. It is incorrect to diagnose ADHD based on these associated problems. Potential for misdiagnosing is very high.

ADHD, like autism, has received quite a bit of publicity recently. The question arises if it is overdiagnosed. Following proper diagnostic procedures, it has been found to occur in 3 to 5% of school-age children. It occurs more frequently in boys than girls. Interesting to note, the same ratio applies when dealing with autism among boys and girls. Experts in the field do not believe that dietary factors cause ADHD. Too bad they have not examined all the adverse effects that a substance like fluoride has on the human brain, especially on the fetus. The Chinese studies have positively shown how fluoride affects the brain in the fetal stage. To this, one can add all the additional toxins that are abundant in our food, air, and water. When I was attending high school, we never heard of anything like autism or ADHD. So what has changed in the past sixty years? **The large number of toxins in our environment that have slowly been released over the years that has eroded our health.**

Children are more vulnerable because they are still growing. Key organs systems such as brain and nervous system, lungs, and reproductive organs are still developing rapidly. The liver and kidneys are not fully developed and can't detoxify. In one study of baby foods sold in the US, more than one-half of all samples contained detectable levels of pesticides. How about the toxic GMO crops present in the baby formulas? Of course, Big Pharma has come to the so-called rescue with a number of synthetic medications. Treating the symptoms of inattention, hyperactivity, and impulsivity, a number of stimulant medications have been introduced. The exact mechanism how it works is not quite known. These medications are prescribed for problems like academic struggles, disruptive behavior, social difficulties, and emotional problems such as depression and/or low self-esteem. The problem with these medications, like all other synthetic medication, is that they come with a list of numerous side effects. In the case of the mental medications, most often the suicidal tendencies are listed.

The following pages are reprinted of an article published by Mike Adams, known as the Health Ranger. He has compiled a very interesting and informative discussion regarding some of the mental medications used.

Reprinted with permission:

Elliot Rodger, like nearly all young killers, was taking psychiatric drugs (Xanax)

Tuesday, June 03, 2014
by Mike Adams, the Health Ranger
Tags: Elliot Rodger, Xanax, psychiatric drugs

> (NaturalNews) Like nearly all mass murderers and psycho killers, Elliot Rodger is now confirmed to have been taking massive doses of psychiatric drugs. Law enforcement authorities have now confirmed Elliot Rodger, the "sorority girl" killer of Isla Vista, California, was taking massive doses of Xanax, a psychiatric drug belonging to a class of chemicals called benzodiazepines.
>
> "Elliot had been taking Xanax for a while, according to his parents . . . there were fears he might have been addicted to it, or taking more than was prescribed," a law enforcement source told RadarOnline (1), which first broke the story. "The Xanax had been prescribed to Elliot by a

family doctor," the story continues. A second story on RadarOnline (2) explores, "disturbing details about the community college student's dependence on Xanax."

That story goes on to report:

Based on interviews with Elliot's parents, Peter and Li Chen, the Santa Barbara Sheriff's Department "is being told that he was likely addicted to Xanax . . . Peter and Li have been doing basic research on addiction to Xanax, and based on what they have read, they believe the tranquilizer made him more withdrawn, lonely, isolated, and anxious," a source told Radar. "It's their understanding that when Xanax is taken in large amounts, or more than the prescribed dosage, these are some of the side effects."

Time after time, mass murderers are found to have been taking psychiatric drugs

Elliot Rodger now joins a long and ever-expanding list of other killers who were either taking psychiatric drugs or withdrawing from them at the time they committed mass murder. While the mainstream media predictably blames guns for all mass shootings, it rarely looks at the chemical drugging of the person who pulled the trigger on those guns. After all, **guns don't operate by themselves**. They require a person to make a decision to commit murder. In case after case, mass murderers on psychotropic drugs describe themselves as feeling withdrawn, isolated, distant and almost living out a "video game" that isn't real. This is what psychiatric drugs do to you: they make you feel detached from reality.

Here's just some of the true history of **psychiatric drugs and mass murder**:

* Eric Harris age 17 (first on Zoloft then Luvox) and Dylan Klebold aged 18 (Columbine school shooting in Littleton,

Colorado), killed 12 students and 1 teacher, and wounded 23 others, before killing themselves. Klebold's medical records have never been made available to the public.

* Jeff Weise, age 16, had been prescribed 60 mg/day of Prozac (three times the average starting dose for adults!) when he shot his grandfather, his grandfather's girlfriend and many fellow students at Red Lake, Minnesota. He then shot himself. 10 dead, 12 wounded.
* Cory Baadsgaard, age 16, Wahluke (Washington state) High School, was on Paxil (which caused him to have hallucinations) when he took a rifle to his high school and held 23 classmates hostage. He has no memory of the event.
* Chris Fetters, age 13, killed his favorite aunt while taking Prozac.
* Christopher Pittman, age 12, murdered both his grandparents while taking Zoloft.
* Mathew Miller, age 13, hung himself in his bedroom closet after taking Zoloft for 6 days.
* Kip Kinkel, age 15, (on Prozac and Ritalin) shot his parents while they slept then went to school and opened fire killing 2 classmates and injuring 22 shortly after beginning Prozac treatment.
* Luke Woodham, age 16 (Prozac) killed his mother and then killed two students, wounding six others.
* A boy in Pocatello, ID (Zoloft) in 1998 had a Zoloft-induced seizure that caused an armed stand off at his school.
* Michael Carneal (Ritalin), age 14, opened fire on students at a high school prayer meeting in West Paducah, Kentucky. Three teenagers were killed, five others were wounded.
* A young man in Huntsville, Alabama (Ritalin) went psychotic chopping up his parents with an ax and also killing one sibling and almost murdering another.

* Andrew Golden, age 11, (Ritalin) and Mitchell Johnson, aged 14, (Ritalin) shot 15 people, killing four students, one teacher, and wounding 10 others.
* TJ Solomon, age 15, (Ritalin) high school student in Conyers, Georgia opened fire on and wounded six of his class mates.
* Rod Mathews, age 14, (Ritalin) beat a classmate to death with a bat.
* James Wilson, age 19, (various psychiatric drugs) from Breenwood, South Carolina, took a .22 caliber revolver into an elementary school killing two young girls, and wounding seven other children and two teachers.
* Elizabeth Bush, age 13, (Paxil) was responsible for a school shooting in Pennsylvania
* Jason Hoffman (Effexor and Celexa) – school shooting in El Cajon, California
* Jarred Viktor, age 15, (Paxil), after five days on Paxil he stabbed his grandmother 61 times.
* Chris Shanahan, age 15 (Paxil) in Rigby, ID who out of the blue killed a woman.
* Jeff Franklin (Prozac and Ritalin), Huntsville, AL, killed his parents as they came home from work using a sledge hammer, hatchet, butcher knife and mechanic's file, then attacked his younger brothers and sister.
* Neal Furrow (Prozac) in LA Jewish school shooting reported to have been court-ordered to be on Prozac along with several other medications.
* Kevin Rider, age 14, was withdrawing from Prozac when he died from a gunshot wound to his head. Initially it was ruled a suicide, but two years later, the investigation into his death was opened as a possible homicide. The prime suspect, also age 14, had been taking Zoloft and other SSRI antidepressants.
* Alex Kim, age 13, hung himself shortly after his Lexapro prescription had been doubled.

* Diane Routhier was prescribed Welbutrin for gallstone problems. Six days later, after suffering many adverse effects of the drug, she shot herself.
* Billy Willkomm, an accomplished wrestler and a University of Florida student, was prescribed Prozac at the age of 17. His family found him dead of suicide – hanging from a tall ladder at the family's Gulf Shore Boulevard home in July 2002.
* Kara Jaye Anne Fuller-Otter, age 12, was on Paxil when she hung herself from a hook in her closet. Kara's parents said, "the damn doctor wouldn't take her off it and I asked him to when we went in on the second visit. I told him I thought she was having some sort of reaction to Paxil . . .")
* Gareth Christian, Vancouver, age 18, was on Paxil when he committed suicide in 2002, (Gareth's father could not accept his son's death and killed himself.)
* Julie Woodward, age 17, was on Zoloft when she hung herself in her family's detached garage.
* Matthew Miller was 13 when he saw a psychiatrist because he was having difficulty at school. The psychiatrist gave him samples of Zoloft. Seven days later his mother found him dead, hanging by a belt from a laundry hook in his closet.
* Kurt Danysh, age 18, and on Prozac, killed his father with a shotgun. He is now behind prison bars, and writes letters, trying to warn the world that SSRI drugs can kill.
* Woody, age 37, committed suicide while in his 5th week of taking Zoloft. Shortly before his death his physician suggested doubling the dose of the drug. He had seen his physician only for insomnia. He had never been depressed, nor did he have any history of any mental illness symptoms.
* A boy from Houston, age 10, shot and killed his father after his Prozac dosage was increased.

* Hammad Memon, age 15, shot and killed a fellow middle school student. He had been diagnosed with ADHD and depression and was taking Zoloft and "other drugs for the conditions."
* Matti Saari, a 22-year-old culinary student, shot and killed 9 students and a teacher, and wounded another student, before killing himself. Saari was taking an SSRI and a benzodiazapine.
* Steven Kazmierczak, age 27, shot and killed five people and wounded 21 others before killing himself in a Northern Illinois University auditorium. According to his girlfriend, he had recently been taking Prozac, Xanax and Ambien. Toxicology results showed that he still had trace amounts of Xanax in his system.
* Finnish gunman Pekka-Eric Auvinen, age 18, had been taking antidepressants before he killed eight people and wounded a dozen more at Jokela High School – then he committed suicide.
* Asa Coon from Cleveland, age 14, shot and wounded four before taking his own life. Court records show Coon was on Trazodone.
* Jon Romano, age 16, on medication for depression, fired a shotgun at a teacher in his New York high school.
  Missing from list . . . 3 of 4 known to have taken these same meds . . .
* What drugs was Jared Lee Loughner on, age 21 . . . killed 6 people and injuring 14 others in Tuscon, Az?
* What drugs was James Eagan Holmes on, age 24 . . . killed 12 people and injuring 59 others in Aurora Colorado?
* What drugs was Jacob Tyler Roberts on, age 22, killed 2 injured 1, Clackamas Or?
* What drugs was Adam Peter Lanza on, age 20, Killed 26 and wounded 2 in Newtown Ct?

Sources for this article include:

(1) http://radaronline.com/exclusives/2014/05/el . . .
(2) http://radaronline.com/exclusives/2014/06/uc . . .
http://www.naturalnews.com/045419_Elliot_Rodger_Xanax_psychiatric_drugs.html#ixzz3E06oCzD0

Chapter 10

Chemtrails

Imagine a world where the skies do not rain milky white or inky black death from the sprayers on specially fitted airplanes. You would have to imagine a world without chemtrails.

Although long denied, governments now admit that chemtrails are *just* weather modification activity, although they are, in reality, one of the mechanisms designed to cut off your genetic line. The genomicidal technologies, first identified by General Bert, are there to make you sick and less fertile and to get rid of you so that the human population will crash. You see, genome disruption is not just what happens to you when you are subjected to GMOs' novel proteins, ionizing radiation from Fukushima, toxic metals from chemtrails and similar sources, vaccines, drugs, and industrial toxins. It is what happens to your genetic present and future if we allow the literally insane genocidal globalist elite to get their way.

Chemtrails are part of an aerial spraying program that appears to be a secret government program or series of programs. Chemtrail poisoning appears to be implicated in Morgellons disease, a horrid condition perhaps caused by the heavy metals, synthetic DNA, GMO viruses, and the other toxins found in chemtrails. Chemtrails are also poisoning our soil, water, and air with arsenic, beryllium, aluminum, and a host of synthetic life-forms. Come with me through the chemtrail door to learn more and learn what you can do to take back the skies and end the destruction of our planet.

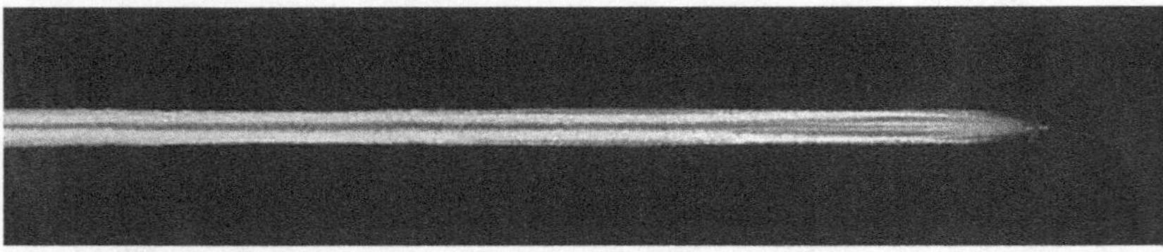

Chemtrails

Another silent killer is chemtrails. You are probably familiar with streaks of white trails after a jet plane flies at an altitude of over more than thirty thousand feet. As the jet fuel is burned by the jet engines, it forms water vapor and quite a few trace amounts of different organic and inorganic compounds. The water vapor at such high altitudes condenses and we can see them as a white tail. Exhaust gases, at high altitudes that jets fly, momentarily condense as ice crystals into thin vapor trails, but quickly dissipate. These are called condensation trails or contrails for short. White tails that do not dissipate within a short time are the so-called chemical trails, or chemtrails for short. Everyone has observed them. Sometimes they can occur on a daily basis crisscrossing in the sky. With time, the white chemtrail expands and almost looks like a white cloud.

Interesting to note that after chemtrails lingered, spread, covered entire skies, and fell, reports of increased illnesses started to be reported, such as persistent coughs, upper respiratory and intestinal distress, pneumonia, lethargy, dizziness, fatigue, disorientation, aching joints and muscles, diarrhea, and nosebleeds. It may affect different people in different ways, but these are some of the symptoms reported. Much of it is reported as flu, but knowledgeable people know that these are poisons dropped on the population. Witnesses have documented and photographed military KC-135s, KC-105s, and white unmarked jets spraying chemtrails. Airport personnel and pilots confirmed specific commercial airliners leaving the long lasting chemtrails. Interestingly enough, it is being reported to the US government, but it is continually denied, that chemtrails do not exist, by the US government. Laboratory examination of these chemicals found that they were manufactured wastes from military industry and bio-warfare substances. In several laboratory analyses, the composition of chemtrails was revealed as containing the following: (1) aluminum barium, (2) aluminum oxide, (3) bacilli and molds, (4) *Pseudomonas aeruginosa*, (5) *Pseudomonas fluorescens*, (6) *Bacillus amyloliquefaciens*, (7) *Streptomyces*, (8) Enterobacteriaceae, (9) *Serratia marcescens*, (10) human white blood cells (a restrictor enzyme used in research labs to snip and combine DNA), (11) *Enterobacter cloacae*, (12) other bacilli and other toxic molds capable of producing heart disease and meningitis as well as acute upper respiratory and gastrointestinal distress, and (13) carcinogen zinc cadmium sulfide.

Chemical researcher by the name of Cliff Carnicom and his associates have documented the poisonous content of the chemtrails by collecting samples via the use of electrostatic devices and HEPA air filters. The analysis of chemtrail fallout has identified many of these toxic chemicals. The list includes aluminum oxide particles, arsenic, bacilli and molds, barium salts, barium titanates, cadmium, calcium, chromium, desiccated human red blood cells, ethylene dibromide, *Enterobacter cloacae*, Enterobacteriaceae, human white blood cells, lead, mercury, methyl aluminum, mold spores, mycoplasma, nano-aluminum-coated fiberglass, nitrogen trifluoride (known as CHAFF), nickel, polymer fibers, *Pseudomonas aeruginosa*, *Pseudomonas fluorescens*, radioactive cesium, radioactive thorium, selenium, *Serratia marcescens*, sharp titanium shards, silver, Streptomyces, strontium, submicron particles (containing live biological matter), unidentified bacteria, uranium, and yellow fungal mycotoxins. This list is provided by the website StopSprayingCalifornia.com.

Let's take a look at some of these pathogens:

Aluminum oxide is toxic to human and animal life. Aluminum builds up in the pineal gland and is linked to Alzheimer's disease and tinnitus (ringing in the ears). Aluminum has a long history of affecting brain functions. Recent research shows that aluminum oxide, which is a heavy metal, can do far more damage to humans and plants. From EPA website: "acid depositions can occur in the wet and dry form and can adversely affect aquatic resources through the acidification of water bodies and watersheds. Aluminum, which is toxic to organisms, is soluble at low pH (acidic medium) and is leached from watershed soils by acidic deposition. Acidification affects fish in several ways, a direct physiological effect of low pH and high aluminum increased fish mortality. The aluminum works with acidic water to create a toxic environment for wildlife in lakes and streams. Plants are affected also." Professor Daniel Watts (a toxicology expert) performed research and reported that aluminum oxide nanoparticles in ground water inhibited the growth of corn, cucumbers, cabbages, carrots, and soybeans.

Earth's natural biological systems are slowly beginning to fail. *Rhizobacteria endomycorrhizae*, a critical microbial organism, is slowly becoming extinct in soils due to barium, aluminum, and heavy metal toxicities. This microbe is responsible for the transfer of nutrients from soils—matrix to the plants/trees feeder-root system. The barium-aluminum heavy metal salts are alkaline and are shifting the base line pH of surface soils and waters to new highs—elevated enough that certain plant species cannot survive.

Ethylene dibromide (EDB), a pesticide and insecticide, banned by EPA in 1984, is a carcinogenic fuel additive. EDB causes central nervous system depression and pulmonary edema, including shortness of breath, anxiety, wheezing, and coughing. It is extremely irritant to mucous membranes and to the respiratory tract.

In the following pages are included some very interesting articles reprinted for your evaluation. One should keep an open mind when reading these reports.

The first article is "Chemtrails – some definitions or when making an observation." Article is from website http://www.holmestead.ca/index-ct.html.

The second article is from a press release in January 2004, an Allison-Wolf report, titled "The Methodic Demise of Natural Earth – An Environmental Impact Overview" Ref. http://www.earthfinaldays.com/methodic_demise_of_natural_earth.htm).

The third article is from an interview with a unnamed scientist who worked on the Shield Program. Since that time, it has been reported that that person supposedly has committed suicide (Ref. http://www.holmstead.ca/chemtrails/shieldproject.html).

Chemtrails—Some Definitions

The word "chemtrail" is not yet in the Oxford English Dictionary and so various terms have come to be used to describe what is seen—such as "anomalous spreading trails" and what the government and others like to refer to as "persistent spreading contrails"—that is without bothering to explain why today plain old-fashioned condensation trails should now behave in this new way!

Contrails have been seen since the beginning of aviation, but those were piston engines burning gasoline and, just like a truck or car on a cold morning, leaving a foggy trail behind.

The chemtrail theory is a group of theories regarding what are claimed to be unnatural condensation trails from high altitude jet aircraft. Contrails are formed by condensation of water vapor in the aircraft's exhausts whereas some trails, or plumes, have an appearance and quality different from those of normal water-based contrails—that is "chemtrails" are not consistent with the known atmospheric properties of contrails.

The term "chemtrail" should not be confused with other forms of aerial spraying such as crop dusting, cloud seeding, military chaff dispersal, or aerial firefighting. It specifically refers to systematic, high-altitude spraying by military-type aircraft of unknown substances for some undisclosed purpose and, in particular, without the informed consent of the air breathing public.

As an aside to the issue of "informed consent": ***Following the recommendations of the Church Committee, President Gerald Ford in 1976 issued the first Executive Order on Intelligence Activities, which, among other things, prohibited "experimentation with drugs on human subjects, except with the informed consent, in writing and witnessed by a disinterested party, of each such human subject" and in accordance with the guidelines issued by the National Commission. Subsequent orders by Presidents Carter and Reagan expanded the directive to apply to any human experimentation.*** **Sources are to be found here: Google search. Now under most recent past and present administrations, those directives have been completely ignored.**

Criminal paid agents attempt to convince all that such chemtrail activity is simply normal traditional condensation trails. In other words, the witness may be mistaken once or twice—but not every time! In fact an accurate, indisputable observation (perhaps with video and/or photographs) on just one occasion is really all that is required to make a case.

Eyewitness testimony is valid in a court of law, so what is the difficulty in accepting it in regards to this illegal spraying?

So here are a few simple, direct points to look for or questions to ask when you see a chemtrail in the sky—and this is in no way an all-inclusive definition list but is simply intended to assist your own personal observations:

- **Is the aircraft in question on regular commercial flight paths or are operating in some restricted, such as military or air space?**
- **Is the aircraft seen on a regular scheduled, perhaps daily, basis or very irregularly—random days and times?**

* **Some areas of a country may have a great deal of air traffic—scheduled passenger, cargo, military, etc.—whereas other areas have very little but still in these areas of very sparse air traffic is the spraying still to be seen?**
* **Is the aircraft seen laying plumes interacting with other similar aircraft in a way that is obviously at odds with normal air traffic such as crossing flight paths and flying at apparently closer than normal spacing or intervals?**
* **Is such trail-creating aircraft seen very frequently, for a period of days, and then not at all for a while in a way that would not be typical of regularly scheduled commercial air traffic?**
* **Is there by chance another aircraft in the sky at a similar high altitude that is apparently on a regular commercial flight path but is leaving behind a short, traditional condensation trail whereas the other aircraft is laying a trail that may stretch from horizon to horizon?**
* **If it is possible to make it out with binoculars, are any heavy trails possibly emanating from areas of the aircraft other than directly behind the jet engine nacelles?**
* **A basic factor in regards to long trails is that the persistent contrail may spread out over the sky and, under certain light conditions, may refract light showing some colors of the rainbow. Has that been seen?**
* **Are the trails laid throughout the sky so that in an hour or so your sky becomes overcast purely from the spreading plumes?**
* **Does the laying of a trail by an aircraft break at times and then start again as if being switched on and off?**
* **If you are well-travelled, do you sometimes visit parts of the world where these trails are never encountered? What could be so special about the atmosphere in that particular country that jet aircraft exhaust behaves so differently?**

- * **Have you by any chance attempted to bring your observations to the attention of the media and then—if there is any response at all—being told something along the lines of *"that is a rather controversial subject"*?**
- * **Have you attempted to learn anything about this illegal spraying from politicians—our so-called representatives?**

This article written by R. Michael Castle, PhD, an environmental professional who holds a National Certification for Environmental Risk Assessment with fifteen years of field practice in environmental risk assessment, investigation, analyses, and remediation. He was a polymer chemist for twenty-two years prior to establishment of the environmental consulting and engineering firm Castle Concepts Consultants Inc.

Dr. R. Michael Castle may be contacted at the following email address: pilot812001@yahoo.com.

The Methodic Demise of Natural Earth

An Environmental Impact Overview

PRESS RELEASE, January 2004, Allison Wolf Report

In this century, we believe we are witnessing the gradual, purposeful demise of the Earth's Natural System. There are those who will debunk/dis-info all that is written regarding the subject of this paper: ChemTrails. What's this? ChemTrails are only a vague description, in lay-terms, of a greater theater of toxic materials being released into the atmosphere/stratosphere, for a myriad of crude and toxic agendas.

The author, Dr. R. Michael Castle, will attempt to put this Global debacle into a profile of events. Technical specificity of all the identified components would require at least a book in length, to recite them all. A short Bibliography follows, and links to various pertinent documents of unquestionable validity.

"First Rule of Understanding: There are very large, winner-take-all games orchestrated by the Global Interests of the World. We have investigated, researched and found substantial evidence of a multiplicity of Global operations designed to mitigate various and theoretical Global catastrophes, as defined by a late 1980's and 1990's International Panel of Conferees. (The value of who are they, what's the politics smell like and all that is not of the utmost importance in our attempt, herein, to describe what serious destruction they have caused and are causing to Humans and our Finite Environment. (Intergovernmental Panel on Climate Change: http://www.ipcc.ch) 1

Many of the operations we have collectively found in our investigations of this debacle have remained a secret, classified and not for general populace information or knowledge. You must answer the question, Dear Reader as to why someone would possibly deploy these Geo-weapons. Just follow the money trails. We will yield all of our voluminous data, information and all references we have used in our validation of facts surrounding this Global atrocity. The recipients of this information must be interested enough in what we are trying to expose, the Human-health risks, the Environmental risks and the plain, simple "wrongness" that prevails. We must determine first if you are provoked to really care, because the research and development of this information for a Major Media whistle-blowing expose will require passion and dedication. Our research work has spanned almost five years of . . . watching and discovering, with disgust, the scenarios which follow.

This is not a small task . . . but the Mission is simple. Expose and Stop the Methodic Demolition of our Natural Earth and its inhabitants.

Executive Summary

Dr. Edward Teller wrote a white paper in the late 1990's describing a remedial operation, strategy, epic-in-proportion, to change the predicted course of what was believed by an International group of Scientists, including Dr. Edward Teller, Livermore National Laboratories, et al, to be the cataclysmic certainty from the results of global-warming, crisis level Ultra-Violet/Cosmic radiation, crisis Ozone-layer depletion and other theoretical doom. (Edward Teller's

"Sun Screen" Document - PDF: http://www.rense.com/general18/scatteringEdTellerwithnotes.pdf)2 Demonstrating here, that the same mind, Dr. Edward Teller, Father H-Bomb, was responsible for many ill-conceived strategies and not one gave considerations to the consequences in the Human realm. Safety, toxicity, lethality, exposure, Environmental Impact, morality, were not words with which Dr. Teller had a high-degree of familiarity. http://tms.physics.lsa.umich.edu/214/other/news/042401HBomb.html)3

These Global-Warming Mitigation Strategies, UV Mitigation and the cessation of the effects of Atmospheric Greenhouse-gases were given a commonality by Teller, et al, and this was the use of a sub-micron particulate. Barium, Aluminum, Thorium, Selenium were to be processed into a sub-micron particle dispensed from high-altitude aircraft, ionized with a specific electrical charge. (BARIUM FLUORIDE OPTICAL CRYSTAL SAFETY DATA SHEET: . (Chemtrails and Barium - absorption and inhalation . . . see links below)(2003 Chemtrails over America (Scroll down to: RADIO FREQUENCY MISSION PLANNER): http://lookupabove.tripod.com/chemtrailsoveramerica)5 We must surmise that ionization keeps the specific heavy metal particulate aloft for longer periods of time. This electronically charged particulate matrix might also be the perfect RF control field. Theoretically, the heavy metals would block and reflect the sunlight from entering the Earth's atmosphere and reflect 1-2% back into space. UV radiation levels would decline. Teller also recommended the use of Commercial Aircraft as-well-as-Military Aircraft to carry out the enormous and epic task of coverage to the Earth's Stratosphere. We believe that the weaponization use of these technologies has been well demonstrated for a US DOD program entitled "RF Dominance." The US Air Force VTRP and the Navy's RFMP are other military programs utilizing aerosolized heavy metal particulates, including aluminized fiberglass or chaffe are characteristic of current military operations. ([2025] Weather as a Force Multiplier: Owning the Weather in 2025: http://www.au.af.mil/au/2025/volume3/chap15/v3c15-1.htm#Disclaimer)6

CIA-led Project Cloverleaf was one of the initial "aerosolized heavy-metal particulate" operations. Massive spraying of the upper-atmosphere/Stratosphere commenced. The US DOD operations soon followed, as the US Air Force became embedded in the overall operations, strategically. The US Air Force would also play significantly into the expansion of a significant Global-Warming Mitigation strategies in the form of Weather Modification and Geo-Engineering practices. (AF2025 v3c15-1 | Weather as a Force Multiplier: Owning . . . | (Ch 1): http://www.au.af.mil/au/2025/volume3/chap15/v3c15-1.htm)7

(Carnicom - Aerosol Operations: http://www.carnicom.com/conright.htm)8

Federally Approved Contractors became involved in this massive, Global effort to save Earth from certain doom. Federal Approved Contractors, FAC's, were part of the research, Development and deployment aspects of these projects, and many of them have visited the premiere website (Stratospheric Welsbach seeding for reduction of global warming: US Patent: 5,003,186-Stratospheric Welsbach Seeding: Click Here)9 (See - Visitors) FAC Academia, multi-National Corporations, US Military/Industrial Complex/Corporations, became eager partners in this effort. One in particular, Hughes Aircraft Corporation of California turned research efforts towards this endeavor.

Thorium and oxides, Aluminum and Silicon carbide have been identified in a special mixture referred to generally as Welsbach Refractory Seeding Agents US Patent 5, 003,186. - March 26, 1991 (Aerosol Programs Patently Obvious: http://users.ev1.net/~seektress/patlist.htm)10 This patent was assigned and awarded in 1990 to Hughes Aircraft Corporation. The Welsbach Refractory Seeding under unrestricted deployment operations began in the early 1990's over a vast area of Stratosphere above the North American Continent. (HAARP HOME PAGE: http://www.haarp.alaska.edu/haarp)11 Expanding over the entire western hemisphere, many operations were believed to have been enjoined in Ozone Depletion Mitigation Aerial operations of the US Air Force, connected to the HAARPs (High Frequency Active Aural Research Project) for Ionospheric Heating, located in Alaska) Ozone Remediation was 1 of 3 active mission parameters for HAARPs, as defined by Dr. Bernard Eastlund, Inventor and Director of HAARPs. The Ionospheric Heating Research Facility was manned and operated by the US Air Force (Reserve) and the US Navy. HAARPs had been weaponized; therefore, all operations were and remain classified. (Correspondence with Dr. Bernard Eastlund: Space Based Weather Control: The Thunderstorm Solar Power Satellite: http://www.borderlands.com/spacewea.htm)12 (Program for Climate Model Diagnosis and Intercomparison: PCMDI Home Page: http://www-pcmdi.llnl.gov)13

Weather Modification/Climate Change

HAARPs is utilized for many clandestine missions, of which Weather Modification is a fundamental objective. Microwave, ELF (Extreme Low Frequency), VLF (Very Low Frequency) and other EMR/EMF-based systems are transmitted into the atmosphere and reflected by the ionosphere, back through the Earth's Stratosphere/Atmosphere where various airborne chemical particulates, polymer filaments and other electromagnetic frequency absorbers and reflectors are used to push or pull the prevailing Jet-Streams, changing Weather patterns. (Note: Vast files of our research are available on the particulate, polymers, Microwave, ELF, VLF and EMG, etc.) In many instances, Drought Inducement Technologies have been found in patented systems. Drought Inducement occurs, according to reviewed technologies by heating the Stratosphere with Microwaves, placing airborne chemical particulates in the airspace and thereby changing the base-line moisture gradients via microwaves from HAARPs and desiccating chemically with Barium Titanates, Methyl Aluminum and Potassium mixtures. (Program for Climate Model Diagnosis and Intercomparison: PCMDI Home Page: http://www-pcmdi.llnl.gov)13" (Is this what is happening on southwest coast presently?)

"HAARPs punches massive holes in the open-air column Ozone, electronically. This is the basis for the Ozone Remediation/Mitigation Strategies found associated with HAARPs. However, "patching" the holes in the Ozone layer has become a standard practice for US Air Force and FAC flight operations. The US Air Force has recently, 2001–2002, resorted to the utilization of Unmanned Aerial Vehicles

(UAV's) technology. High-flying Stratospheric Robotic Platforms negate the manned operation factors. Robots don't complain and never talk and remain, forever, non-union. Welsbach Seeding and Ozone Hole Remediation sciences utilize chemistries that are toxic to Humans and the Environment. Welsbach Refractory Agents use Aluminum, Thorium, Zirconium and other emissitivity-refractive metals and metals oxides. Thorium is elementally, 98% purity. The remaining 2% are identified radioactive materials. Ground-fall includes Thorium. Mid and Eastern Canada are suffering from what has been clinically identified as Thorium poisoning. There are no other base-line resources for Thorium, all from residuals of aerosolized heavy metal particulate spraying into the Stratosphere.

Alaska Flight 261

One other observation has been made regarding Welsbach refractory agents—the extreme abrasion characteristics of some of the other patented components, namely, Aluminum Oxide and Silicon Carbide. These materials are extremely high MOH's Hardness and only second in abrasativity to Diamond. These 1 micron and sub-micron dusts, upon falling through the Atmosphere, could conceivably find deposition onto a somewhat, adhesive surface, inside the working flight components of all aircraft traveling through the "grit-plume." The greases used on the Horizontal and Vertical Stabilizers, Ailerons, Flaps and Landing Gear trucks may be seriously damaged with abrasion of metals on the aircraft. A horizontal-stabilizer Jack-Screw continuously coated with these highly abrasive dusts from the Welsbach Refractory materials will cause a gradual milling of the jack-screw metals and cause complete failure, jamming the flight controls into an uncontrollable down or up attitude configuration. We believe that Alaska Airlines Flight 261 was a victim of this unforeseen circumstance. Alaska Flight 261 made daily passage through heavy grit plumes from ChemTrails operations associated with Welsbach Refractory Seeding operations, principally along the West Coast of the US, down throughout Dallas, Texas. Other aircraft, Commercial and Military, Civilian have also suffered these flight-component failures and attributed them to sub-standard Aviation Machinist's quality. As did the NTSB (National Transportation Safety Board) in the Alaska Flight 261 aviation fatalities. These conclusions could not have been farther from the Truth in these matters. Collateral damage or just another consequence of ChemTrails?

Ozone-Hole Remediation—The composition most notably utilized in Ozone-hole patching is elemental Selenium and an Aromatic Hydrocarbon such as Toluene (Benzene component of Gasoline) and mixed isomers of Xylene. Sprayed from Stratospheric flying Aircraft, films of this toxic mixture fall into the area just above the Troupopause, the Ozone Layer. Ozone or triatomic Oxygen forms rapidly upon the irradiation of the Selenium and the Hydrocarbon with Ultra-Violet/Actinic sunlight. This is the identical photon/chemical reaction that causes Ozone Alert Days and is problematic. The solid-state reaction of Selenium and Ultra-violet radiation is the same as the reaction that occurs during Xerography. Copy machines generate minute amounts of Ozone when the Selenium Toners are irradiated with UV light sources. The US/NASA TOMS (Total Ozone Mapping) Satellites have verified these Ozone Patching operations that occur principally over the northern sectors of the North American Continent. We have been studying these phenomena since the early 2000-time line. (TOMS—Total Ozone Mapping Spectrometer: http://toms.gsfc.nasa.gov)14

Ground-fall Impacts of Selenium/Aromatic Hydrocarbons—The aromatic Hydrocarbon, when ground fall occurs, (there is substantial evidence that this has repeatedly occurred in the United States and Canada) is causative of Benzene overexposure. Carcinogenicity of Benzene is known, not suspect. Surface water pollution with Benzene is a continuous threat.

Selenium poisoning is characteristic of most heavy metals including Lead (Pb).

Weather Modification—An integral part of the US Department of Energy's (US DOE) Global-Warming/UV Mitigation/Climate Change, Strategies. Led by the US Air Force, as a distinct Geo-Weapons method, Weather Modification is occurring on a Global basis. (AIR FORCE 2025 FINAL REPORT: http://www.au.af.mil/au/2025)15

This Executive Summary will be confined to ChemTrails deployment in the United States and Canada due to the expansive nature of a global theatre. Professional independent investigators have retrieved numerous samples of soils and waters; these were analyzed and documented. Extensive testing has demonstrated the highly suspicious toxic levels of Barium and Methyl Aluminum in many areas of the United States and Canada that have no extraneous sources of these heavy-metals. The sample retrieval areas have been documented for extreme ChemTrails observation. Accompanied by verbal and written complaints to many Public Health Care Agencies, causes for the elevated levels of Barium and Aluminum found in the bloodstream were not forthcoming; just another clinical conundrum. (See Attachment - Chemtrails Pertinent Links)

Conclusions—Although this summary is only a fraction of the overall ChemTrails debacle, we encourage impartial investigators and/or credible news journalists to assess this core information and either validate or refute our findings, thus far. The entirety of the Human Health suffering and the Environmental Impacts are staggering and too voluminous to write at this time. Other aerosol-related programs will be discussed at a follow-up session. Moreover, biological materials and genetically mutated fungi used as bio-controls may even be more damaging and egregious to Humans or the Environment than chemical-based aerosolized chemical materials sprayed into the atmosphere. (RIZZO DM - Oak Tree Sudden Death: http://www.plpnem.ucdavis.edu/plp/people/bio_info/Rizzo.htm)16

All of the chemical spraying operations have been conducted and deployed without public knowledge and not one Human Health Risk Assessment nor Environmental Impact Study has been submitted for Public and Civil scrutiny or for comments. This alone is an egregious violation of Civil, Environmental and International Laws, with respect to Treaties. The Natural Earth's biological systems are slowly beginning to fail. Rhizobacteria-endomycorrhizae, a critical microbial organism, is slowly becoming extinct in soils due to Barium and Aluminum heavy-metal toxicities. This microbe is responsible for the transfer of nutrients from soils-matrix to the plants/trees feeder-root system. The Barium/Aluminum heavy-metal salts are alkaline, i.e. Barium Hydroxide, and are shifting the base-line pH of surface soils and waters to new highs—elevated enough that certain plant species cannot survive. According to our experts in this field, without this microbe, natural plant growth is impossible. The ChemTrails are spoiling our Infinite Natural Eco-System and no one is watching, albeit the evidence for microbial extinction events abound.

The events of the 1990s were characterized with phrases such as; "re-inventing" Commerce, Government, etc., are uniquely defined by the clandestine operations carried forth by the embedded personnel of US Agencies, in collaboration with International Alliances with subjects such as Weather Control, Food and Water Control and expansion of pharmaceuticals treating new diseases, all having direct linkage to what we have compressed into something that is called ChemTrails. Does this story interest you in the least? **We, a group of concerned individuals from around the World, numbering in the millions, are now extremely convinced that we are witnessing the methodic demise of the Natural Earth in the name of Commerce and Control, big Corporate and a hideous form of Socialism via the Military Industrial Complex. The technologies identified in ChemTrails are crude, poisonous and lethal. The bankrupt independent Farmer of the United States and Canada never knew that the Drought that had-no-end and destroyed their businesses was actually contrived. Seems though, all their properties were purchased for pennies-on-the-dollar by either International Farming interests or International Grain-Farming interests and all that is grown and harvested on these farms are genetically mutated grains and foodstuff, most likely with a Monsanto Patented Product. These farms do not take USDA Subsidies either. Well, that's good for the US Taxpayer, right? This is the emergence of the Food Weapon.**

Our ultimate mission is to develop Legislation for the United States Congress to pass that will halt, prohibit and forever abate the practices of deploying, dispensing or otherwise releasing any chemical into the Atmosphere/Stratosphere of the airspace above the entire North American Continent. Such a Draft Law has been written by the Author, R. M. Castle, entitled: The Unified Atmospheric Preservation Act of 2003 (UAPA3), has been reviewed by other ChemTrails Investigators and seems to be the only method of intervention to halt these insidious methodologies. However, the US Congressman we hoped would carry our cause, tends to back off when it becomes clear that the US Government is strategically involved in many aspects of the debacle. We are not asking for this legislation to be considered, we will demand its passage. Once in force, who will then become bankrupt? We will shut down the HAARPs Electromagnetic radiation machine with UAPA3. I would be very honored if anyone would like to receive a copy of the draft legislation UAPA3, for review and/or endorsement.

Dr. Castle may be contacted at the following email address: pilot1981@voyager.net

References:

Intergovernmental Panel on Climate Change: http://www.ipcc.ch
Edward Teller's "Sun Screen" Document - PDF: http://www.rense.com/general18/scatteringEd-Tellerwithnotes.pdf

Edward Teller's "sun screen" document "Global Warming and Ice Ages: I. Prospects for Physics-Based Modulation of Global Change," E. Teller et al., August 15, 1997, Lawrence Livermore National Laboratory, as PDF, presented at the twenty-second International Seminar on Planetary Emergencies, Erice (Sicily), Italy

Who is Edward Teller?: http://tms.physics.lsa.umich.edu/214/other/news/042401HBomb.html

This is the same Dr. Edward Teller known as Father-H-Bomb. (The same person that once recommended "nuking" a prominent East Coast US harbor in order to make it deeper for ships of international flags' commerce. In the Super-Port that would result, international commerce and trade would be exceptional, especially one that would not be a union-controlled port. Nonetheless, someone nixed that idea after they reviewed the potential lethality of the residual radiation for at least one thousand years.

1. Barium Fluoride Optical Crystal Safety Data Sheet: http://www.crystran.co.uk/baf2safe.htm
2. 2003 Chemtrails over America (Scroll down to: RADIO FREQUENCY MISSION PLANNER): http://lookup-above.tripod.com/chemtrailsoveramerica
3. [2025] Weather as a Force Multiplier: Owning the Weather in 2025: http://www.au.af.mil/au/2025/volume3/chap15/v3c15-1.htm#Disclaimer

For ease in downloading above paper:

PDF: Weather as a Force Multiplier - Owning the Weather 2025: http://www.au.af.mil/au/2025/volume3/chap15/vol3ch15.pdf

1. AF2025 v3c15-1 | Weather as a Force Multiplier: Owning . . . | (Ch 1): http://www.au.af.mil/au/2025/volume3/chap15/v3c15-1.htm
2. Carnicom—Aerosol Operations: http://www.carnicom.com/conright.htm
3. Stratospheric Welsbach seeding for reduction of global warming: US Patent: 5,003,186-Stratospheric Welsbach Seeding: Click Here
4. Aerosol Programs Patently Obvious: http://users.ev1.net/~seektress/patlist.htm

Aerosol-related links:
http://www.aaar.org/hplinks.htm

1. HAARP HOME PAGE: http://www.haarp.alaska.edu/haarp
2. Correspondence with Dr. Bernard Eastlund: Space Based Weather Control: The Thunderstorm Solar Power Satellite: http://www.borderlands.com/spacewea.htm
3. Program for Climate Model Diagnosis and Intercomparison: PCMDI Home Page: http://www-pcmdi.llnl.gov
4. TOMS—Total Ozone Mapping Spectrometer: http://toms.gsfc.nasa.gov
5. AIR FORCE 2025 FINAL REPORT: http://www.au.af.mil/au/2025
6. RIZZO DM - Oak Tree Sudden Death: http://www.plpnem.ucdavis.edu/plp/people/bio_info/Rizzo.htm

The material presented here is by R. Michael Castle, PhD, Castle Concepts Consultants Inc. You may reprint this information with the author's permission.

Here we quote the communication from Deep Shield:

Points to Ponder: The Shield Project

Reports of biological material being part of the spray should be addressed. Therefore, I will give as much attention to all of your questions as possible.

1. What purpose do polymer threads imbedded with biological material serve in this scenario?

   Polymers are part of the mixture and they do form in threads and in `tufts'. The idea is simple and comes to us from the spider. As you may know spider webbing is very light, some newborn spiders spin a `parachute' to catch the prevailing breeze to travel far from their place of birth. Spiders have been able to attain high altitudes and travel great distances for long periods of time. Most of the elements used in the spray are heavier than air, even in their powdered form they are heavier and will sink quickly. Mixing them with the polymers suspends the particles in the atmosphere high above the surface for longer periods of time, therefore in theory we do not need to spray as often or as much material. Since the suspended particles eventually do settle into the lowest part of the atmosphere and are inhaled by all life forms on the surface there is an attempt to counter the growth of mold by adding to the mixture mold growth suppressants—some of which may be of biological material.

Mold comes in spores that travel on the winds; the polymers can attract **mold** spores through static charges created by the friction of the polymer threads and the atmosphere. Add a bit of warmth and moisture and mold begins to grow. The polymer is stored in a liquid form as two separate chemicals. When sprayed they combine behind the plane `spinning' long polymer chains (threads). Much tinkering has been done which the chemical matrix in past years. Many polymers (plastics) are non-biodegradable thus add to the problem of pollution. Various formula have been used, some which even use biological agents. It would be great if we could reproduce the same web material that spiders make, it is extremely strong, extremely lightweight and breaks down relatively fast in the ecology.

2. If this spraying is to mitigate global warming, why does so much of it take place at night?

Though it would appear that the dispersal rate of the spray is fast, it is actually takes much longer to be an effective shield. There is a desired concentration being sought. One that is thick enough to stem the UV and the

Infrared, while being thin enough to allow visible light through. A perpetual cloud cover would have disastrous effects on plant life; the food chain thus disrupted would soon collapse. The desired effect wanted is a thin cover that would theoretically create a daytime haze that allows plenty of sunlight while providing protection. From UV radiation and also reflect enough infrared to maintain nominal temperatures.

The optimal condition is to use the least amount of material to provide the maximum amount of shielding. Ideally that would be a one-time application which would stay suspended for years, however, as noted, barium and aluminum and other trace elements are far heavier than air and they sink rather rapidly. The different temperatures between day and night causes massive volumes of air to rise during the night, the warm air trapped at the surface rises above the cooling air above. By strategically spraying in certain areas at night, we get the advantage of the rising air, which not only pushes the material higher, but also causes the material to disperse into a thin layer.

I would suggest studying on the subject of weather, namely highs and lows and how air moves to fully understand the times of spraying. I note, it is not just global warming we are combating here, we are also combating UV Summer. Global warming could effectively be treated by applications during the night, when the warm air rises. However the UV needs to be treated during the day. This is why on some days one finds that more spraying is done during the day. The UV indexes are monitored constantly for local areas. If the problem were simply cooling the earth, rockets would have been used to suspend particles in the high atmosphere. However the delicate nature of the Ozone Layer precludes this method of shielding. More on this in the answer to Question 6.

3. What other military programs are in place involving the spraying of barium and what are their purposes? Do you know and understand the chemical make-up of the element?

A little knowledge will go a long way to understanding the need to use barium: Barium is often used in barium-nickel alloys for spark-plug electrodes and in vacuum tubes as a drying and oxygen-removing agent. Barium oxidizes in air, and it reacts vigorously with water to form the hydroxide, liberating hydrogen. In moist air it may spontaneously ignite. It burns in air to form the peroxide, which produces hydrogen peroxide when treated with water. Barium reacts with almost all of the nonmetals; all of its water-soluble and acid-soluble compounds are poisonous. Barium carbonate is used in glass, as a pottery glaze, and as a rat poison. Chrome yellow (barium chromate) is used as a paint pigment and in safety matches. The chlorate and nitrate are used in pyrotechnics to provide a green color. Barium oxide strongly absorbs carbon dioxide and water; it is used as a drying agent. Barium chloride is used in medicinal preparations and as a water softener. Barium sulfide phosphoresces after exposure to light; it is sometimes used as a paint pigment. Barite, the sulfate ore, has many industrial uses. Because barium sulfate is virtually insoluble in water and acids, it can be used to coat the alimentary tract to increase the contrast for X-ray photography without being absorbed by the body and poisoning the subject.

Note what Barium Oxide can do, absorb carbon dioxide—one of the chief gasses causing the green-house effect. In my answer to Question 4 I will discuss the need to carry a current in the shield. I would like to point out that barium and aluminum work together to diffuse and strengthen an electrical charge. Somewhat like the current produced when acid is introduced between two dissimilar metals, such as iron and copper. There are military applications for everything you can think of, can not a butter knife be used as a weapon? The same concept holds true here.

4. What is the connection between ELF, EMF, VLF and Chemtrails spraying? Or is there one?

To understand the use of radio waves in the shield, one first understands how ozone is created. I cannot stress to you how dire the situation really is. The shield in place is only a partial solution; we must counter the depletion of the ozone—this means we must make ozone in the stratosphere. Ozone at ground levels does no good; indeed, ozone pollution at ground levels it what is used to determine the air quality. Higher levels of ground level ozone mean that air quality is bad. Pure ozone is an unstable, faintly bluish gas with a characteristic fresh, penetrating odor. The gas has a density of 2.144 grams per liter at standard temperature and pressure. Below its boiling point (-112?) ozone is a dark blue liquid; below its melting point (-193?) it is a blue-black crystalline solid. Ozone is triatomic oxygen, O3, and has a molecular weight of 47.9982 atomic mass units (amu). It is the most chemically active form of oxygen. It is formed in the ozone layer of the stratosphere by the action of solar ultraviolet light on oxygen. Although it is present in this layer only to an extent of about 10 parts per million, ozone is important because its formation prevents most ultraviolet and other high-energy radiation, which is harmful to life, from penetrating to the earth's surface. Ultraviolet light is absorbed when its strikes an ozone molecule; the molecule is split into atomic and diatomic oxygen: 03+ ultraviolet light—>0+02. Later, in the presence of a catalyst, the atomic and diatomic oxygen reunite to form ozone.

Ozone is also formed when an electric discharge passes through air; for example, it is formed by lightning and by some electric motors and generators. Ozone is produced commercially by passing dry air between two concentric-tube or plate electrodes connected to an alternating high voltage; this is called the silent electric discharge method. Since UV radiation is the problem, we cannot use UV to produce more stratospheric ozone. Another method must be found. The shield acts like one plate of the electrode, when tickled with certain radio waves; it produces an opposite charge to stratospheric layers producing low atmosphere to stratosphere lightening. Creating ozone where it is needed.

5. If this is being done for the reasons you say, then why are other chemicals being used, why are different sprays being used?

   Correcting the ecological damage that mankind has done has NEVER BEEN DONE BEFORE. We are relatively new to this notion of terraforming on a real scale. That is what we are doing, Terraforming. We are trying to recreate the ideal life-sustaining conditions on a dying planet. We have never done this before, not intentionally. We are testing and trying different methods. Granted, if we do nothing 89% of all species will go extinct and humanity stands a high chance of not surviving through two more generations (or less). However the idea of 2 billion casualties death and permanent injury is not easy to swallow either.

   Several attempts to improve the application of Shielding material and getting the most out of each application are taking place all the time. The combined resources of the nations of earth are not enough to allow constant spraying. Though we have achieved a high level of technology, there is a great surface area that needs to be covered nearly daily. Large sections of ocean are all but ignored; the remaining land masses are more than what can be covered effectively. The Shield would work best if it was a single thin layer without interruption, however due to the movement of air, weather patterns and the sad fact that we do not have the means to place ample amounts of material at the same level at the same time we are getting a small fraction of the effectiveness from our applications.

6. Why is spraying found before storm fronts? Is it to cause drought?

    Before a storm there is a front, the front clears the air before a storm, pushing particulate matter ahead of it, leaving a space relatively clear of particulate matter. UV radiation levels rise in these areas, sometimes to dangerous levels. The shield must be maintained. Since barium absorbs water as well as carbon dioxide, precipitation has been affected. Other kinds of sprays are in development and testing which may reduce the affects on precipitation. As I stated above, this is a new technology we are working with, it is still in its infancy and there are some problems with it.

7. Why are UFO's and disappearing spray planes reported?
   I do not know.

8. What about the reports of sickness after spraying?

    There are several causatives for this. Some people are more sensitive to metals, whiles others are sensitive to the polymer chemicals. As stated in a previous email, people will get sick, and some will die. It is estimated that 2 billion worldwide will be affected to some degree by the spraying. Without spraying we have a 90% + chance of becoming extinct as a species within the next 20 years.

9. What is the relationship between these spraying programs and One World Order?

    Personally I am against the move for globalization, and yes, there is potential to use the Shield to speed up the process of globalization, there are several countries that are involved in this project: European Union Nations, USA and Russia are the largest contributors to the project, many of the allied nations and UN Members participate to one extent or another. The material (chemical spray as you may call it) comes from all of these nations.

To insure that the chemicals are not tampered with, they are mixed and sprayed over random nations. This means that chemicals produced in the USA has a good chance of being sprayed over Russia, England and the USA. This random spray of material means that no nation would be certain that their chemicals will be sprayed over a nation which they have issues with. Russian planes may be seen in USA skies, but so too will USA planes be seen in Russian skies. The canisters used are sealed in a third nation that has no idea where its canister is going. Participating nations have their observers at every station where canister loading is done. All of this to insure that the shield is not used as a weapon. To further insure that the shield is not used as a weapon, non participant nations are sprayed by participants who must spray in order to get enough material to maintain their nations shield. It is understood that not spraying is as much a military offense as shooting at them. Without the shield, UV poisoning would cause great death. The threat is a common one, which has brought nations together in defense. The natural outcome of having a common enemy is to strengthen international ties—a step toward globalization.

10. Is the Spraying related to terrorism?

Yes and no. Recent terrorist activity can be traced to resistance groups who feel that we should not interfere with the natural order of things. As you know, there are many rumors out there as to what the Shield Program is. Some believe that this is a population reduction scheme, designed to kill off 'undesirable' peoples. While others hold that this is a mind control program. There are many theories which have sinister plots in them these are propagated by the resistance groups in an attempt to stop the shield regardless of the consequences.

The same delivery method could be used for biological and chemical warfare. It could also be used to inoculate large populations, the effectiveness of these uses are low, there are better methods that can be used. As a means to fight terrorism it is ineffectual, it is far easier to inoculate a population individually and would insure full inoculation against germ warfare.

11. Why all the secrecy?

Due to the severity of the situation it is mandatory to maintain public calm for as long as possible. The Earth is dying. Humanity is on the road to extinction—without the Shield mankind will die off with in 20 to 50 years. Most people alive today could live to see this extinction take place. This means that an announcement of the situation we face boils down to telling every man, woman and child on earth that they have no future, they are going to be killed. People would panic. There would be economic collapse, the production and movement of goods would collapse. Millions would die in all cities on earth, riots and violence would reduce civilian centers to rubble within days. Half of the population in dense metropolitan areas would try to leave the cities seeking 'safety' in the rural areas thinking that they would be safe. Those left behind in the cities would be at war with their neighbors, fighting for the remaining supplies. We would be telling the world that the world is coming to an end, and even with the Shield the chances of survival are small.

UV Summer and Global Warming are the immediate problems we face, there are far greater problems that are raising their ugly heads and will present new problems which in some cases have no viable solutions at this time. Ecologies are collapsing. The extinction rate of species is climbing. The amount of chemical pollutants in the water and soil are fast approaching and in many places has surpassed the earth's ability to heal itself. Crop failure is on the rise, even in the USA the returns on crops are smaller than they were 10 years ago. Even with the advances in genetically altered food crops, we are falling behind in our ability to produce enough to go around. Throughout the 20th century chemical fertilizers and pesticides were used to insure the best yields. Unfortunately many of these have contaminated ground water, killed beneficial insects along with the undesirable insects. These chemicals have gotten into the food chain and are affected other species besides mankind. It is only a matter of years before famine spreads like a cancer throughout the world.

Clean fresh water is in short supply, in many places well water is non-potable, containing the run off of pesticides, herbicides and fertilizers that have been used on crops and lawns. The water treatment facilities we have are unable to scrub out all of the toxins we have placed in the soil and water supply. Many of the toxins we find build up over time in the body, a long slow poisoning which has been making its presence felt in many areas of the world in the form of cancers, leukemia, sterility, birth defects, learning disorders, immune deficiency problems, etc. These are on the rise, any good researcher can find the records. For decades there was public outcry for the end to pollution. For every small step we made to clean up our production, millions where born who added to the problem. Yes, pollution is down per individual, however there are a couple more billion individuals producing pollution, thus the real numbers have an increase in over all pollution produced. Name a city that does not have problems with smog. You would be hard pressed to find one. Though smog controls on automobiles is higher than ever before, the number of autos on the road has increased thus the amount of smog producing pollutants is higher than ever before. All the clean air acts passed to curb individual factory and auto emissions did not address the production of more factories and more autos. Here an uneasy compromise was made between the need to maintain the economy against the need to maintain the ecology. The ecology lost since it was estimated to be a problem decades from now. The economy was a problem that would have dire effects today.

All of these factors combined have produced a scenario that in shorts boils down to the end of the world in 50 to 75 years. Even if we were to stop all emissions of pollution today, the inertia of past decades is enough to carry us over the brink in 100 years. However we cannot stop the production of pollution, to do so would mean shutting down every factory, every auto, every train, truck, ship and every household on the planet. Electricity is used to heat many homes in the Western World. The production of electricity produces fewer pollutants than heating all homes with wood or coal. Cutting our power generation abilities down to hydroelectric and fission reactors would leave a good chunk of the world in the dark. It is an impossible situation, our civilization is geared to the use of energy, take away our energy and civilization will collapse.

12. When will spraying stop?
There are several factors governing this:

A. Should the Ozone layer repair itself or our active attempts at repair reduces the amount of ground level UV to acceptable levels, spraying will stop. Present calculations place this between 2018 and 2024.

B. Should another method be found which is more effective, less costly or presents us with long-term solutions the Shield Project would be replaced.

C. When the other problems become too big to make the maintenance of the shield worth the effort. The estimated date for this is 2025 to 2050.

13. Since Global Warming and UV summer are the problem, why is the Government backing down on its pollution controls?

Because they are ineffectual and will cause more economic problems than they would solve ecological problems. We surpassed the threshold of Earth's ability to absorb pollutants in the 1970's. Since that time the earth's population has nearly doubled. Emerging Industrial nations have come into being, more pollutants are produced now than back then, even with the stringent controls in place. The world is heading for economic depression, more emission controls would add to the economic problems. This translates into our being unable to do anything to start solving the problems.

Unfortunately our technologies require a strong economy to advance. We need that advancement, we need the trillions of dollars spent on research that a strong economy causes. Each corporation that produces a product has a product development program in place. Many of the past products invented came by accident through other unrelated products. There is a corporate drive to find methods to clean up the ecology, to reduce emissions, etc. These goals have been in place for decades, many of the large corporations are in the know when it comes to the ecological problems we face thus they are spending a great deal of money and time on finding solutions to the problems we face. Take away the economy and their research stops.

14. How are you related to the Chemtrails? How do you know that this is what is happening?

I would prefer to not state who I am or how I am related to all of this. To validate what I say, would require a bit of research on your behalf. I would recommend the following subjects to look up and study:

A. Population numbers for industrial nations and the tons of pollutants produced annually. Start with 1975 and work your way up.
B. Number of emerging Industrial Nations.
C. Number of cases of Skin cancers worldwide.
D. Crop Production vs. land area dedicated to crop production. Simple math will show that more acreage is needed to produce food per individual.
E. Automobile production from 1975 to present, estimated number of autos on the road and the average emissions of later model cars produced as compared the emissions of earlier model cars. A little math will show that though individual autos produce less emissions, the amount of emissions has risen due to the number of autos on the road. Remember that many autos are the road that were built before present emission control standards. 1980 is a cut off date—anything put on the road before then produces more pollutants than autos produced today. I would include research in the number of diesel autos produced, diesel has not been under the emissions control acts.
F. Severity of storms and the number of severe storms. Also include heat waves and droughts in that research, you will find that the numbers are staggering when compared to data from 1950, 1960 and 1970.
G. Research how naturally occurring Ozone is produced in the stratosphere. Compare to how it is produced industrially.

H. Research political reforms in the past 30 years, see which political institutions have changed, which nations have joined with whom. Concentrate more on these from 1982 onward. This would include the fall of the Wall and Iron Curtain.

I. Research polymers and how they are made, look at recent research done in biological polymers, medical polymers and filaments.

J. Check out spiders and spider web and the way spiders use their different webs and threads.

K. Research clean fresh water estimates as compared to the 1970's to today—world wide.

L. Research the following medical conditions per capita: Birth Defects Cancers Leukemia Immune deficiency diseases (excluding virus borne ID illnesses such as HIV) Occurrences of Learning disabilities, including dyslexia, ADD, and over all IQ tests Sterility for both male and females world wide Instances of glaucoma and cataracts.

M. Compare the history of UV indexes from 1970 to present. You may note that it was on sharp rise until 1997-99.

N. I would strongly recommend researching the reactions of different barium and aluminum compounds and how they are used. Research how long it takes for these metals in pure form to oxidize, how they combine with nitrates, carbon monoxide carbon dioxide and fluorocarbons and hydrocarbons and water vapor.

O. Research how mold propagates, the conditions it needs to grow and just how abundant it is in the atmosphere.

If you pursue these lines of inquiry, you will see the Shield Project as it really is

At the time of the US-lead invasion of Iraq I had the opportunity to ask a few supplementary questions. There has been no attempt to integrate these questions and answers into the previous section therefore some may appear somewhat out of logical order.

A couple of the questions have a Canadian approach. By the way, I came up with the name—have to call the source something!

Here we quote the further communication from "Deep Shield":

15. Could you, "Deep Shield"—be described as a scientist or . . . ?

    Scientist is a good generic term. I do study and research in a scientific manner. I carry papers and degrees. My official capacity is in direct research of atmospheric issues in relation to pollutants. I also create models of potential long-term effects of green house gasses on the climate. Predict wind patterns, weather patterns, etc.

    I have spent a good many years working on the project calculating the amount of material needed and creating models for dispersion patterns. I work other members who know the chemicals used and their interactions with the atmosphere, pollution and water vapor. I am part of a team which itself is part of a larger team, which is part of still a larger team. Government work with many chiefs and levels above the workers.

16. Are you prepared to comment upon your personal motives for sharing this information?
    Not at this time, no

17. Have you signed a non-disclosure or Secrets Act document that specifically relates to this project?
    Yes.

18. To your knowledge what categories of individuals officially know of the Project and are expected to remain silent? For

example, my list of suspects includes government down to the county level, military especially air force, meteorologists, health specialists, mainstream media etc.

All those who know are expected to remain silent. All of those who suspect are either faced with trying to prove the virtually unprovable or are faced with good enough reasons to remain silent. I would assume that this situation is worldwide and could be considered one of the dangers of this project.

It was presented to me as a matter of national security. I can see the reasons why there is a desire to repress the information not that spraying is taking place but the hard little fact that we are facing a period of human history which might be the end of civilization.

19. Is the mainstream controlled media specifically ordered to avoid any mention of chemtrails? If so, have you anything further to add such as how was this done?

I would assume that the Media is controlled by its own desire to make money from what it reports. Since there is enough debunking out there, which says that contrails are part of the normal use of jet engines in the atmosphere, this would leave a reporter with very little to report unless there was solid evidence or pictures or something that could not be explained away.

You must know by now all the debunking methods that have been employed. The 'official' announcements are the media's main dish. The rest they regulate to the realm of the National Enquirer.

20. What government agency or agencies control this program? Is it under international control?

It is an international program. Many nations contribute in different ways. Measures have been taken to insure that what is sprayed over all countries is the same through triple blind deliveries; which include not knowing where a certain canister will end up, not knowing which aircraft a certain canister will be flown, and not even knowing who (in military craft) will be piloting a craft which has the purpose of spraying (Note: in today's world there is usually a mixed crew of different nationalities flying any one military aircraft on a Shield mission). I believe the Media caught Canadians in Iraq recently when Canada's official say on the matter was that Canada was not giving any support to the military might.

The fact remains that there were Canadian military with the USA forces. Some on aircraft carriers most being pilots. I think you can connect the dots.

21. How is the project funded—who pays for it? Have you any idea of the total direct operating cost? Also, does Canada make a funding contribution for the activities in our skies?

    Most governments tend to over charge themselves to cover for their black operations (unofficial operations). That money comes out of the collection of taxes. So in effect the taxpayers of the world are paying for this project.

    I would assume Canada does contribute funding to the Project. Canada is one of the top nations contributing time, material and funding to this project. Most of the Free World, the Western World, has taken on most of the burden of the costs.

22. Is the Shield Project the only such aerial spraying program?

    Is it the only project designed to avert ecological disaster? Then yes. There are countless other projects that could be taking place which include spraying of some sort or another. Pesticides are usually sprayed. There has been great interest in weather control such as bringing rain to arid regions and taking the punch out of hurricanes and typhoons.

Weather control may be one of the final options left to us. Considering the amount of global warming that has taken place. There is a strong need to deflect a storm's fury, or to bring rain back to those regions which have been suffering drought.

What Mother Nature has done for millions of years automatically may now require mankind's hand to keep the schedule.

23. There are reports of four different chemtrail programs and other "code" names. For example, see: Holmestead: Chemtrails - what are they? Any comments?

It is possible that the Military does have a use for similar sprays. I cannot speak for the Military. However, my own personal research has come across these things as well. Are they possible? Yes. Are they practical? Only in the small scale say over the battlefield, or in the case of say the Iraq War, over Baghdad. Global application would be far too expensive and would require an obvious flight pattern of grids, circles and other heavy spray patterns.

24. Is all the spraying done using the "tank kits" described earlier or are the KC-135R and KC-10 types filled to the brim? Such aircraft have a load capacity of 200,000 pounds or more for refueling missions.

No. Several types of craft are used. Commercial jet airliners are used and they are not diverted from their flight paths to do so. How the canisters and the spraying is done on this kind of craft is unknown to me exactly. I do have my suspicions. I know best that which is my field; this is not to say that we do not talk around the water tank. So I know more than just my area and am able to think the matter through to its logical end.

I do know that even all the commercial jetliners in use are not enough to insure complete coverage all of the time. My computer models require knowing how much material needs to be sprayed. Certain conditions cause wide areas to suddenly (over hours) open up in the Shield. Then and only then is mass spraying done—and would be done with the most logical craft, a tanker.

Why not spray more from individual jetliners? That is one of the problems. Jetliners do not carry much material (100 to 500 gallons) because the material has to be spread out thinly.

Look at the kinds of material being used, aluminum, barium, titanium, etc. Most are highly reflective; in some instances the material is an absorber of gasses. In the case of reflection the desire is to reflect X amount of heat and X amount of UV while still maintaining acceptable (nominal) levels of UV and heat reaching the planet's surface.

Life requires a certain amount of both UV and Heat too much will kill—so will too little. The apparent amount looks like a lot more than what is actually being sprayed per volume of air it is covering. Most of the whitening of the sky is not the material per se, but the collection of water vapor, which forms into suspended ice crystals. The introduction of the material causes the water vapor to collect like rain collects on individual particles of dust.

Too much material would cause a "mud fall" of sorts where the naturally occurring water vapor would precipitate carrying the material with it.

Spraying is done in such away as to "layer" the material through a volume that will allow an acceptable level of UV and heat through along with all the other wavelengths of light. Photosynthesis is the foundation of life on our planet.

Only when all the material is removed in a local area does it require a massive spray, this is usually in the front of a weather system, or after a heavy period of precipitation. Then a tanker is flown, fully loaded.

25. Is there any truth in the story that some of the spraying is done by jetliners with modifications in the "honey" or waste compartment? For example, see mechanic story: Mechanic.

The technology used for spraying is rather simple. It requires at least two tanks under pressure, each carries half of the mixture which is sprayed at the same time forming a complete compound which is designed to be lightweight (so as to be suspended for longer periods of time).

There have been attempts to incorporate the materials in jet fuel, however the material binds with unburned jet fuel, water vapor, etc and does not have the added buoyancy of the polymer threads. The end result is a spray that is less than half as effective and is more dangerous since it can lead to sulfates, acids and other mixtures, which are more lethal than the spray.

It is very possible that the "honey" compartment is used. The amount of material needed is small compared to the payload of any given commercial airliner.

However, there is a good deal of fuel tank that is not used. Airliners only fuel their craft for the journey ahead of them; they rarely top off the tank. This has become public knowledge in light of 9-11. It was this small fact that caused the terrorists to pick pan-continental flights so they would have a plane fully loaded.

The majority of flights are short range and do not require the full capacity of an airliners fuel tanks. Any adaptations needed could easily be done during routine maintenance, and could be easily explained away as being a modification for safety and-or pollution controls.

This last is my own theory.

We can assume that any means possible to deliver the material is tried. Independent nations may favor one way of doing so over another.

26. Where are the official sources that state that a certain number of people (worldwide?) will sicken and possibly die as a result of the spraying? In other words, what *internal* studies have been done on the health issues and who carried them out?

WHO (World Health Organization) carried out most of the studies. Other nations have carried their own research on the matter. Some have said the ill effects will be minimal—along the lines of a million or so, while others have found the numbers to be far higher—3 to 4 billion.

Some of the organizations include the CDC and independent labs. We are dealing with a situation where the amount of contamination is estimated to be far higher than what would normally take place but is far lower than historical instances of industrial contamination. This is important to note, the only real history we have with barium/aluminum/titanium etc. contamination is through factory workers, miners, etc, who receive a far greater dosage of the material than what is to be experienced by the populace under the Shield.

The amount of spray is very small compared to the volume of the space that is covered. Most of the harmful chemicals that are used are being dissipated over vast areas. Near coastal regions the fall out is not reaching land at all, but is being carried out to the oceans. The addition of polymers to make the material remain suspended in the air longer means that less material is being used.

Today the material used and its application is nothing like in the early days when it was sprayed in greater quantities and settling far faster to be inhaled by all.

The accepted Estimated Casualties (from WHO) is 2 billion over the course of 6 decades. The majority will be either the elderly, or those who are prone to respiratory problems. These numbers are based on the current estimates of the general health of the population, the average

age and the occurrence of respiratory problems as a health issue. All are estimates since there are no solid numbers to work with.

27. Could you summarize the root causes of the initial destruction of the atmosphere that requires this "repair" work? Did it perhaps result in part from fluorides released/produced by the nuclear weapons programs?

In a word—Industry. Most fail to understand that the products we use, wear and live with are made in a manner that dumps CFC's and green houses gasses into the atmosphere. There is no one single causative in this issue. It goes way back to the Industrial Revolution and the use of coal to power steam engines. Since that time we have consumed greater and greater energy resources, dumping the waste where ever we wanted.

Up until very recently refrigeration was a big contributor, imagine all those hundreds of millions of households that owned and operated freon cooled refrigerators from 1940 to 1970. Not just one refrigerator per household, but over the course of time often multiple freon units. This doesn't include the various air conditioner systems or industrial refrigeration systems.

For a long period when the refrigerator or air conditioner unit was replaced, the old one was taken to the dump and thrown into the heap—the freon was free to escape and make its way up into the stratosphere to eat away at the ozone layer

You can add to that list. Think of all the cars that had air conditioners, think of all those hair spray cans with their propellant gasses—the amount of those alone were enough to do great damage.

Styrofoam is another industry and product that has contributed to the problem. In the scheme of things atomic energy has contributed little compared to the consumer goods that have been manufactured during the past century.

Think of all the cars on the road today. In the late 1970's smog controls started getting stronger. Think of all the pre-1978 cars on the road—they are still producing a good many chemicals that leech into every corner of the globe.

28. Have you any comment on the use of so-called "CloudBusters" based upon the theories of Wilhelm Reich? For an example of this see: CloudBuster.

    I could go on for hours about the particular errors of this site. (*15 September 2005: Note that due to misunderstanding of this answer by some it is necessary to make clear that this reference by DS is to the: CloudBuster website mentioned in the question.)* Let us just cut to the chase: I seriously doubt that this machine does as it is reported to do. I could be wrong, but then I do not work with alternative energy sources such as Orgone energy.

Here is the second item mentioned above:

Points to Ponder: Conroy Penner
British Columbia

Conroy Penner, of Victoria, British Columbia, Canada contacted me directly with his story. He told me that some years ago he used to work for a certain company in Western Canada and along with another person who was a qualified aerospace welder/fabricator he was assigned to work on construction of some special spray equipment for the United States Air Force.

The workers were told that the USAF contract involved equipment to be used to spray insects at airforce bases in the States. The spray systems incorporated exotic alloys and specially machined parts with large holding, pressure and surge tanks along with pumps.

Penner became suspicious of the actual purpose of the whole project as, in his opinion, things simply did not "add up." He resigned after there were confrontations with the US military people and the management over the true purpose of the equipment

Penner provided this photograph, that was taken in 1988 of some finished spray equipment on a flatbed trailer. There are parts carts in the foreground.

The description states that the green tank sections are coated aluminum and the others are stainless steel for certain other chemicals. The smaller tanks are for the solution for purging the system. The large tanks are some 15 feet long and approximately 3 feet in diameter. The box seen at the right side of the photograph is the control panel, and the plumbing and pumps are underneath the tanks. It is said that these units were designed for use with Hercules C-130 aircraft and it is understood that in total eleven systems were to be made, this being the first. I have on hand many other specifics from Penner but that is the general outline. All of this took place in 1988/89.

Comments: There are obviously some problems with this—it would be great to have the exact dates along with the names of the company management involved. Also the names and rank of the US military personnel involved is lacking.

As for the equipment itself—it looks much as if it could be ordinary aerial spraying systems—that is because indeed it may be just that with suitable modifications for the specialized USAF requirements. At this time the aerial spraying programme (chemtrails) was in the early stages of experimental development. Certainly at first it was reported that the C-130 was seen being used for aerial spraying (of unknown substances) and only later were the large refueling tankers seen spraying "chemtrails."

Further comments: Added 26 March 2003—is an addition from (and confirmed by) the source of the main item above. The canisters that are shown above are similar to what is known to be incorporated into this programme. Two large ones are carried, one carrying one part of the chemical mix, the other carrying the other part, when combined they form long polymer chains—threads if you will. Even the green tint finish is typical. However, the way it is understood here, is that the small tank provides the propellant (compressed gas) which is released into the larger tank. Much like a large aerosol can, except the compressed gas is stored in a separate tank which is much stronger, and able to sustain a far greater pressure than a larger tank. Further confirmation of this particular aerial spray system is found at this United States Air Force Reserve, 910th Airlift Wing of Youngstown, Ohio website: Air Force Reserve - MASS. This page will open in a separate window and may be kept open, minimized or closed to return here. Here the first page is quoted in part: "Six Modular Aerial Spray Systems (MASS) are available at the 910 AW to conduct aerial spray missions.

Each system can be configured with up to four 500 gallon tanks for a total volume of 2000 gallons.

USAF C-130 MODULAR AERIAL SPRAY SYSTEM (MASS)

The Aircraft. Four C-130H aircraft are modified to perform aerial application. Modifications consist mostly of ULV (Ultra-Low Volume) and LV (Low Volume) wing line installation and electrical modifications.

Spray System. The MASS is built by Lockheed/Conair. One of the major design criteria of the MASS was that it had to be a "roll-on/roll-off" system allowing the aircraft to be reconfigured for spray or airlift in under an hour. To accommodate the Roll-on/Roll-off design, the full MASS is designed in 3 modules, each attached to modified standard (463L) aircraft cargo pallets. The operators console, pumps, catwalks, and cradles for flush and chemical tanks are all secured to these modified pallets. The pallets lock into the C-130s dual rail system. Once the MASS is loaded, interconnecting plumbing and electrical circuits tie the MASS modules together. To contain any spillage of spray materials, a 1.5" lip (drip pan) surrounds the pallets. The dry weight of the MASS is ca. 10,500 lbs."

The 910 Airlift Wing webpage continues with more technical and operational details.

Holmestead.ca does not necessarily endorse the views expressed in this webpage. We present this in the interests of research and for the relevant information we believe it contains relating to the illegal high altitude spraying by large military type aircraft of unknown substances—commonly referred to as chemtrails

The big impression made on the interviewer was that the planet is dying, and that there is an increase in Ultraviolet light radiation due to global warming. If that is the case, then why do chemtrails contain so many other adverse ingredients such as metals, biological material, desiccated blood and God knows what else.

Subsequently returning to the Holmstead index page, it was discovered that the person giving this anonymous interview has died. It was reported that that person has committed suicide. How convenient and coincidental. Maybe too much information was given out regarding chemtrails, and their potential application for weather manipulation or military application. It is also interesting to note, that the chemtrails are only applied over densly populated areas. One country does not know the sprayed contents, since it was packaged in a different country according to the report given. The main two ingredients mentioned are aluminum nd barium particles, plus polymers. From previous pages we have learned what these do to our environment, including some of the bacteria and other derogatory ingredients.

An interesting thought occurred to me. When we first came to Detroit, Michigan, the furthest north that a tornado came was Monroe, Michigan, which is about 30 miles south of Detroit, Michigan. Now, in September 2014, there are reports of potential tornadoes in the Upper Peninsula of Michigan. I remember that the tornadoes were not as frequent in the '50s, '60s, or '70s as they are now. The same can be said about hurricanes. Now the weather channel reports that we can expect any place from five to 15 major hurricanes in a season. Could this be a result of the spraying that started in the 1980s, 1990s, and continuing to present day? Also, I never heard of any tornadoes in Poland for a number of decades, until recently. Is this part of geoengineering to use weather as a weapon? You can access GeoengineeringWatch.org to learn more.

Major Media Articles on Chemtrails and Geoengineering

Chemtrails: Is US Gov't. Secretly Testing Americans
November 9, 2007
KSLA-TV (CBS affiliate in Shreveport,
http://www.ksla.com/Global/story.asp?S=7339345

Could a strange substance found by an Ark-La-Tex man be part of secret government testing program? That's the question at the heart of a phenomenon called "Chemtrails." "It seemed like some mornings it was just crisscrossing the whole sky. It was just like a giant checkerboard," described Bill Nichols. He snapped several photos of the strange clouds from his home in Stamps, in southwest Arkansas. Nichols said these unusual clouds begin as normal contrails from a jet engine. But unlike normal contrails, these do not fade away. Soon after a recent episode he saw particles in the air. "We'd see it drop to the ground in a haze," added Nichols. He then noticed the material collecting on the ground. "This is water and stuff that I collected in bowls," said Nichols as he handed us a mason jar . . . after driving down from Arkansas. KSLA News 12 had the sample tested at a lab. The results: A high level of barium, 6.8 parts per million, (ppm). That's more than three times the toxic level set by the Environmental Protection Agency, or EPA. We discovered during our investigation that barium is a hallmark of other chemtrail testing. This phenomenon even attracted the attention of a Los Angeles network affiliate, which aired a report entitled, http://www.youtube.com/watch?v=4RABRI8q4EI"target="_blank">ToxicSky? There's already no shortage of unclassified weather modification programs by the government. But those who fear chemtrails could be secret biological and chemical testing on the public point to the 1977 US Senate hearings which confirmed 239 populated areas had been contaminated with biological agents between 1949 and 1969. Later, the 1994 Rockefeller Report concluded hundreds of thousands of military personnel were also subjected to secret biological experiments over the last 60-years. http://www.ksla.com/Global/story.asp?S=7339345"

This article has been removed from the KSLA website, but is https://web.archive.org/web/20070505141027/http://www.ksla.com/Global/story.asp?S=7339345 is still available on the Internet Archive. This is one of the few media reports to break a veritable wall of silence in the major media on the disturbing chemtrail phenomenon. To watch this revealing newscast on YouTube,

Years Ago, The Military Sprayed Germs on US Cities, October 22, 2001, Wall Street Journal.

http://online.wsj.com/article/SB1003703226697496080.html

http://online.wsj.com/article/SB1003703226697496080.html

Much of what the Pentagon knows about the effects of bacterial attacks on cities came from . . . secret tests conducted on San Francisco and other American cities from the 1940s through the 1960s, experts say. In the 1950s, Army researchers dispersed Serratia on Panama City, Fla., and Key West, Fla., with no known illnesses resulting. They also released fluorescent compounds over Minnesota and other Midwestern states to see how far they would spread in the atmosphere. The particles of zinc-cadmium-sulfide—now a known cancer-causing agent—were detected more than 1,000 miles away in New York state, the Army told the Senate hearings. In New York, military researchers in 1966 spread Bacillus subtilis variant Niger, also believed to be harmless, in the subway system by dropping lightbulbs filled with the bacteria onto tracks in stations in midtown Manhattan. The bacteria were carried for miles throughout the subway system. The Army kept the biological-warfare tests secret until word of them was leaked to the press in the 1970s. Between 1949 and 1969 . . . open-air tests of biological agents were conducted 239 times, according to the Army's testimony in 1977 before the Senate's subcommittee on health. In 80 of those experiments, the Army said it used live bacteria that its researchers at the time thought were harmless. Several medical experts have since claimed that an untold number of people may have gotten sick as a result of the germ tests. These researchers say even benign agents can mutate into unpredictable pathogens once exposed to the elements. Reading the full article on the Wall Street Journal website requires a paid subscription. You can read it free at http://www.mindfully.org/Reform/

Military-Germs-US-Cities.htm." Considering that the army kept all of these tests secret for decades, is it possible that they are keeping information on chemtrails secret from the public?

Special Report: 'Toxic Sky?" Home May 24, 2006, NBC4, Los Angeles

http://www.nbc4.tv/news/9155725/detail.html"

http://www.nbc4.tv/news/9155725/detail.html</SPAN></A></P>

It's a quiet mountain community, but some residents claim something's happening in the sky that's making them sick. Mystery clouds and unusual contrails . . . Is it a weather experiment on a massive scale? In a Channel 4 News investigation, Paul Moyer looks into why some say the government is manipulating the weather. Watch: http://web.archive.org/web/20060615141713/www.nbc4.tv/video/9265818/detail.html"

Video Report References: http://commerce.senate.gov/hearings/witnesslist.cfm?id=1683"

US Senate Committee testimony on Weather Modification http://csat.au.af.mil/2025/volume3/vol3ch15.pdf" Owning the Weather in 2025

http://www.agriculturedefensecoalition.org/content/california-skywatch"

California Skywatch (Rosalind Peterson https://web.archive.org/web/20060627062622/http://www.alpenhornnews.com/modules.php?name=News&file=article&sid=450" Alpenhorn News Stories This webpage on chemtrails and geoengineering was taken down by NBC4 for some reason. Yet you can still view it thanks to the Internet archive at http://web.archive.org/web/20060618094715/http://www.nbc4.tv/news/9155725/detail.html" Below are excerpts from the two parts of this investigative news report with links to view the original broadcasts online. The webpage at the above link to the US Senate Committee testimony has also been removed, but unfortunately is not available on the Internet archive. May 24, 2006, NBC4, Los Angeles http://www.youtube.com/watch?v=JNDMJCTFslw http://www.youtube.com/watch?v=-JNDMJCTFslw</SPAN></A></P> Is the government experimenting with our weather? Chemtrails: The Air Force says there is no such thing. On the Internet they are sighted as proof of the government creating clouds to combat global warming. It is called geoengineering; fighting global warming by putting a chemical dust in the atmosphere and reflecting harmful radiation back into space. You can use barium oxide, for example, which makes big fluffy clouds. You could use tiny little bits of aluminum, which is benign in the environment, and essentially manage the climate. Rosalind Peterson of http://www.californiaskywatch.com" California Skywatch Health Department records show a sharp increase in both chemicals in the water supply of Northern California, dating from the time the huge contrails first appeared at her home over in Mendocino County. "What I found was unusual spiking since the early 1990s in barium, aluminum [and more]. All these things in the same test would be up way over state and federal standards. This had to be airborne, because how could it get to such diverse regions of our county." A bill soon to be voted on in the US Senate—a weather modification act—[calls] for research in "attempting to change or control by artificial methods the natural development of atmospheric cloud forms. "All of this information on chemtrails was removed from the NBC4 website for some reason, yet you can still find some of the original information using the Internet archive at

http://web.archive.org/web/20060618094715/http://www.nbc4.tv/news/9155725/detail.html

And thankfully, the original broadcast is available on YouTube at the link below the title and date of this summary September 2006, NBC4, Los Angeles http://www.youtube.com/watch?v=Ev8AMhkesuo, More now on a Channel 4 news investigation . . . concerning some strange clouds in the skies over Southern California. Some say they are simply jet contrails. Others . . . call them chemtrails, an alleged government program to combat global warming by creating clouds with chemicals. Former FDA analyst Rosalind Peterson says the evidence on the ground only raises more questions about what's in the air. "Every time they put a chemical in the air . . . we're starting to see an unusual spiking in California's drinking water supplies." She's searched through public documents . . . from NASA and the

Air Force and found a long history of testing in the atmosphere with chemicals like barium and aluminum, said by the agencies to be used at safe levels. Peterson . . . says she combed over drinking water data from the last 10 years available from the state department of health. Her conclusion, those same chemicals; barium, and aluminum and others, are showing up in water supplies across the state spiking sometimes above standards set by the department of health and often around the same time as atmospheric test dates. The EPA tells us that high levels of aluminum and barium can affect the nervous system and cause high blood pressure. All of this information on chemtrails was removed from the NBC4 website for some reason, yet you can still find some of the original information using the Internet archive at http://web.archive.org/web/20060618094715/http://www.nbc4.tv/news/9155725/detail.html, And thankfully, the original broadcast is available on YouTube at the link below the title and date of this summary. Chemtrails Are Over Las Vegas August 19, 2005, Las Vegas Tribune http://web.archive.org/web/20051124041624/http://www.lasvegastribune.com/20050819/headline1.html"

http://www.lasvegastribune.com/20050819/headline1.html</SPAN></A> (Part 1 http://web.archive.org/web/20051025134500/http://www.lasvegastribune.com/20050826/headline3.html" http://www.lasvegastribune.com/20050826/headline3.html</A>

(Part 2) Las Vegas residents are increasingly noticing the appearance of chemical trails overhead. Such "chemtrails" are substantially different in appearance to the normal condensation trails left by jet airliners. The difference is that while condensation trails are composed of water vapor that dissipates rapidly, "chemtrails" linger much longer and spread out over time to eventually cover the sky with a thin haze. The US Air Force Website refutes the "Chemtrail Hoax"

as having been around since 1996. Before you believe . . . the government's "denial," do an Internet search for the following terms: http://www.google.com/search?q=%22Joint+Vision+for+2020%22 Joint Vision for 2020

and http://www.google.com/search?num=100&hl=en&newwindow=1&q=%22Weather+is+a+Force+Multiplier%3A+Owning+the+Weather+in+2025%22Weather is a Force Multiplier:

Owning the Weather in 2025, a whitepaper by MIT's Bernard Eastlund and H-bomb father Edward Teller. Before he died in 2003, Teller was director emeritus of Lawrence Livermore National Laboratory, where plans for nuclear, biological and directed energy weapons are crafted. In 1997, Teller publicly outlined his proposal to use aircraft to scatter through the stratosphere millions of tons of electrically-conductive metallic materials, ostensibly to reduce global warming. Two scientists working at Wright Patterson Air Force Base confirmed . . . that they were involved in aerial spraying experiments. In the US Air Force research study, "Weather as a Force Multiplier" issued in August 1996, seven US military officers outlined how http://www.wanttoknow.info/war/haarp_weather_modification_electromagnetic_warfare_weapons HAARP and aerial cloud-seeding from tankers could allow US aerospace forces to "own the weather" by the year 2025. Among the desired objectives were "Storm Enhancement," "Storm Modification" and "Drought Inducement."

The Las Vegas Tribune is not a leading newspaper, yet this is one of the few significant media articles on the chemtrails and geo-engineering. For more from a good alternative website, http://weatherwars.info/?page_id=12. Is it just a coincidence that the writer of this article served as managing editor of the Tribune? http://web.archive.org/web/20060207063858/http://lasvegastribune.com/20051104/headline3.html" Marcus Dalton, was fired on Oct. 21st, two months after this article was published? Army test in 1950 may have changed microbial ecology October 31, 2004, San Francisco Chronicle (San Francisco's leading http://www.sfgate.com/health/article/Serratia-has-dark-history-in-region-Army-test-2677623.php, http://www.sfgate.com/health/article/Serratia-has-dark-history-in-region-Army-test-2677623.php, Serratia is a bacterium that some doctors and residents of the Bay Area have been familiar with for many years. In 1950, government officials believed that serratia did not cause disease. That belief was later used as a justification for a secret post-World War II Army experiment that became a notorious disaster tale about the microbe. The Army used serratia to test whether enemy agents could launch a biological warfare attack on a port city such as San Francisco from a location miles offshore. For six days in late

September 1950, a small military vessel near San Francisco sprayed a huge cloud of serratia particles into the air while the weather favored dispersal. Army tests showed that the bacterial cloud had exposed hundreds of thousands of people in a broad swath of Bay Area communities. Soon after the spraying, 11 people came down with hard-to-treat infections at the old Stanford University Hospital in San Francisco. By November, one man had died. The outbreak was so unusual that the Stanford doctors wrote it up for a medical journal. But the medics and [the dead man's relatives didn't find out about the Army experiment for nearly 26 years, when a series of secret military experiments came to light. Some people now speculate that descendants of the Army germs are still causing infections here today. The secret bio-warfare test might have permanently changed the microbial ecology of the region. The military regularly used humans as guinea pigs in experiments in the decades before and after WWII. For a list of these sometimes lethal experiments, http://www.wanttoknow.info/humanguineapigs." For reliable information on government mind control experiments which also used unsuspecting civilians, http://www.insightcourse.net/lessons/14a_mind_control." Considering that the army kept all of these tests secret for decades, is it possible that they are keeping information on chemtrails secret from the public? Read below for what you can do to make a difference. For an informative panel discussion on chemtrails and geoengineering which was held at the prestigious Commonwealth Club in San Francisco in March 2011, http://commonwealthclub.org/events/archive/podcast/man-made-climate-change-skies-32811." And if you still question the reality of chemtrails, consider that the military regularly used humans as guinea pigs in experiments in the decades before and after WWII."

What you can do: Inform your media and political representatives of this critical information on chemtrails and geoengineering. To contact those close to you, http://www.wanttoknow.info/contact-mediapoliticalrepresentatives" Urge them to bring this information to light and allow public dialog on the topic of chemtrails. To read an excellent, well researched article on the related field of HAARP, electromagnetic warfare and secret weapons, go to http://www.wanttoknow.info/war/haarp_weather_modification_electromagnetic_warfare_weapons" Learn about the intriguing history and development of controversial behavior modification programs in this excellent article http://www.wanttoknow.info/mind-control, Footnotes and links to reliable sources are provided for verification purposes.

Index of chemtrail pages at Holmestead

This page is an index of the many pages at Holmestead devoted to the chemtrail debate. There are also other sub-pages and documents not directly listed here—plus external links. Please note that this is not a news site or a blog and although some material may be dated it is still valid. There is no single point of view or theory being presented here but a variety of views and some may indeed at first appear to contradict others. Also note the cautions or disclaimers on a few pages. These pages will automatically open in a separate window or you can Right Click and open in a New Tab or Window.

Chemtrail aircraft spraying

- ** Petition January 2013: A formal Petition to be presented to the Canadian House of Commons regarding "chemtrails."
- ** "Definitions": Some definitions or at least some points to consider when making observations of suspected "chemtrails."
- * Points to Ponder - comments: A collection of material, related links and other resources.
- * Chemtrails - our first experience: Photographs and descriptions of some of our first observations starting in Spring 2002.
- * Additional photographs - 1: More photographs and descriptions from 2002.

* Additional photographs - 2: More photographs and descriptions—please note all these images were from around the Holmestead.
* Holmestead map: Where we are in the World.
* Chemtrails - what are they?: Some educated guesses from 2001.
* Government correspondence: Asking questions of various Canadian governments and their responses plus face to face meetings with local representatives.
* Access to Information Act: Access to Information requests proved that Canadians from coast to coast were prodding governments for answers.
* "Whistle-blowers"?: This is the well-known "Deep Shield" interviews plus other such material.
* Jim Phelps: Comments on the Shield project.
* Radar and Soil Tests: Spray Tankers Tracked by Radar, Lab Tests Raise Concerns.
* Holmestead soil tests: Soil tests of Holmestead area farm fields.
* Aluminum and your health: Aluminum is in chemtrails—how aluminum may affect health.
* Petition in English and French: We went through the process of placing an official petition before the Canadian Parliament.
* Le Goût de Vivre: Le journal local Le Goût de Vivre a édité l'information dans deux éditions pour la communauté franco-ontarienne.
* "Fact-filled" article: This article was published in a local newspaper.
* NavCan: Some information was obtained from NavCan, the private Canadian ATC company—plus details of the death of a pilot.
* Strategies Against Climate Change: A good discussion for anyone looking more deeply into the subject.
* Weather-mongers: About the folks who make a living telling us what is going on in our skies.

* State Secrets - State Lies: This page explores why you should not have high expectations of hearing the truth from your government.
* Election issue?: An activist attempted to make chemtrails a local election issue.
* Europe - Raum+Zeit: An excellent report published in Europe.
* Europe - Raum+Zeit on Greenpeace: The same journal as above but this time reporting on how Greenpeace ignores the "chemtrail" issue.
* The Solar connection: The July 2004 issue of National Geographic magazine plus a related report.
* Orb report: A personal report of a chemtrail spraying "orb" over Lake Ontario.
* CT calling cards: These are useful to hand out to people to take away with them and perhaps decorate their fridge.
* Print a chemtrail poster: An idea for a poster that you may want to consider.
* Digital altitude + airspeed: Using a digital camera to calculate altitude and speed of aircraft.
* Climate Change Jekylls and Hydes - pdf: How scientists are approaching the "geoengineering" required for climate control - by Wayne Hall.
* Pilots' view: Some links to photographs taken from high altitude aircraft.
* Spring 2005: A variety of topics bringing us more up to date.
* Fall 2005: Another update with a collection of material from near and far.
* Weather Modification Bill: Rosalind Peterson is very active in Redwood Valley, California and here draws attention to the pending US Bills.

Plus there is more.

* Climate Change and Climate Modification - pdf: Some of the politics surrounding the climate change debate - by Wayne Hall.
* KNBC - "Toxic Sky": KNBC in Los Angeles, broadcast this investigative "chemtrail" report 23 May 2006.
* Discussion: Wayne Hall and Rosalind Peterson—a dialogue on persistent jet contrails and experimental weather modification.
* Spreading Trail Identification System: A method of making positive identification of the plumes by Louis Aubuchont of Parsonsfield, Maine (USA).
* Discovery Channel: Television series, Best Evidence, "Chemical Contrails" episode and "Weather Warfare" a new book.
* Global Warming: Dr. Timothy Ball with some cold, hard facts.

The Holmestead is located at: 1, 17th c40 Thunder Beach Road Concession,
Township of Tiny, Ontario, L9M 0T3, Canada
Other very excellent source is GeoengineeringWatch.org.

Chapter 11

GMO

GMO stands for "genetically modified organism." Additional acronyms are GM for genetically modified or GE for genetically engineered. Biotechnology in agriculture has come to mean the process of intentionally making a copy of a gene for a desired trait in one plant or organism and using it in another plant. GMOs are plants or animals that have been genetically engineered with DNA from bacteria, viruses, or other plants and animals, in ways that cannot occur naturally. DNAs are the ribonucleic acids that contain genes, which give the respective plant or animal its specific characteristics. All plants and animals contain DNA. Genetics is the study of genes.

The introduction of foreign germ plasm into crops has been achieved by traditional means by overcoming fertility barriers. A hybrid cereal was created in 1875 by crossing wheat and rye. Since then some of the traits introduced into wheat were dwarfing genes and rust resistance.

Genetic modification is a result of an organism's manipulation by inserting new genetic information into existing cells in order to modify a specific organism for the purpose of changing its characteristics. New DNA may be inserted by first isolating and copying the genetic material of interest using molecular cloning methods to generate a DNA sequence or by synthesizing the DNA, and then inserting this into the host organism. Genes can be removed using nuclease. Nuclease is any group of enzymes that split nucleic acids into nucleotides and other products. Gene targeting is another technique that uses homologous recombination to change an endogenous gene, delete a gene, remove exons, add a gene, or introduce point mutations. An organism produced through genetic engineering is considered a genetically modified organism, or GMO.

Genetic engineering has been applied in numerous fields such as research, agriculture, industrial biotechnology, and medicine. As an example, enzymes used in laundry detergent and medicines such as insulin and human growth hormone are now manufactured in GM cells, experimental GM cell lines, and GM animals, such as mice, for research purposes. Genetically modified crops have been commercialized.

Biolistics is also a method of genetically modifying food. It is usually used in plants, by shooting new genes into the host. Microscopic particles of gold or titanium are coated in the DNA sections which are to be introduced to the host. These are fired into plant cells and the rest depends on these particles entering the cell's nuclei. Gene silencing is a technique that doesn't add a gene but rather removes one. First, the trait to be removed is identified, then another copy of the gene is attached, but in opposite direction, therefore essentially cancelling out the desired trait.

http://www.disabled-world.com/fitness/gm-foods.php

http://www.ruralsource.ac.nz/TempDocuments/Resources/Genetically

ModifiedFoodFA3.pdf

Starting in 1996, these **genetically modified organisms, or GMOs,** spread rapidly through US fields. Following is a graph representing the percent of acres planted in the US for five genetically modified crops: HT soybeans (dark blue), HT cotton (red), Bt cotton (light green), Bt corn (lilac), and HT corn (light blue).

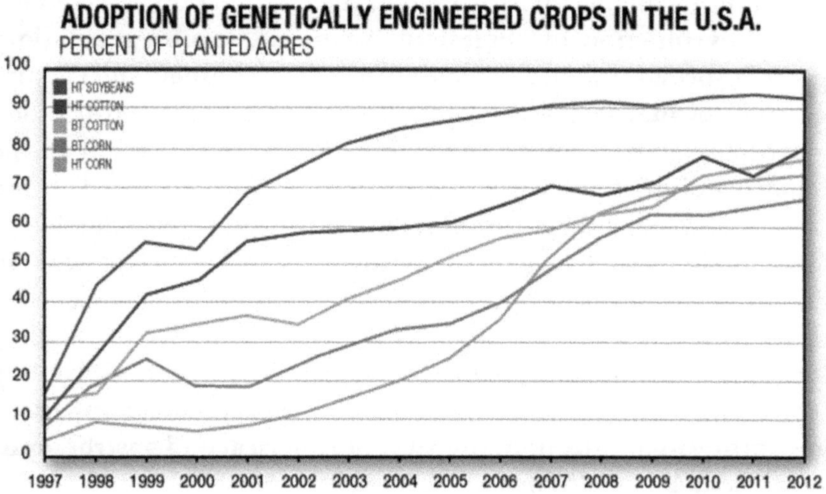

Genetically engineered seeds are patentable; therefore, the companies that develop these seeds can have total control of the market. This makes the companies very profitable. A company like Monsanto, as an example, controls 90% of the market. They also are the predominant company supplying the necessary insecticides, pesticides, and herbicides that are applied to these GMO crops.

Farmers who purchase Monsanto's patented GMO seeds have to sign a contract. This agreement forbids farmers from keeping any new seed produced by the crop for the following year. They cannot even save leftover seeds. Use of the seeds is one time only. This ensures a perpetual stream of cash flow for the company since farmers will have to buy new seeds every year. If the farmers should use leftover seeds or gather seeds from their crops, they face a lawsuit. The other problem encountered in the farming industry is that the GMO pollen infects adjoining farms using conventionally farmed seed crops. Once this cross-pollination has been established, the farmers are sued for patent infringement. This is adding insult to injury. Our representatives in Washington, DC, have added a rider to one of their laws passed since, that Monsanto cannot be sued. Research from American Academy of Environmental Medicine states that "several animal studies indicate serious health risks associated with GMO food consumption including **infertility, immune dysfunction, accelerated aging, dysfunction of cholesterol synthesis, insulin regulation, protein formation, and changes in the liver, kidney, spleen, and gastrointestinal system."** Many scientists believe that GMO food is what's triggering inflammation, which can in turn result in the following diseases: coronary artery disease, arthritis, cancer, diabetes, obesity, and many other chronic ailments. Since the introduction of GMO food, there has been an increase in a number of ailments in our population in the United States. A reporting agency by the name of Laissez Faire Club has compiled very interesting graphs regarding some of the more common illnesses reported to date. According to the club, the four common illnesses are **chronic constipation, gastrointestinal infection, diabetes in America, and gastroesophageal Reflux. http://www,laiseszfaireclub.com**.

A major percentage of our immune system estimated at about 80% resides in our intestines or gut. It is our principal defensive system against pathogens. The friendly flora, bacteria that reside in our gut, not only provide us with beneficial nutrients and vitamins but also produce antibacterial compounds lethal to the pathogens. If the gut senses food as a foreign item, which is the case with GMO foods, it will attack it, creating an inflammatory response. Inflammation flares up whenever our body feels threatened. As the below graphs clearly indicate, when the GMO foods were introduced into our food supply, the gastrointestinal problems have increased tremendously. But it does not stop there, since the damage to our health travels beyond our digestive tract. Recent medical research indicates that hidden inflammation in our gut is the root cause in many chronic illnesses, such as heart disease, obesity, diabetes, depression, cancer, and even autism.

The secret killers in our food cause the following chronic conditions represented graphically over the years prior to and after GMO crop release on the US market:

CHRONIC CONSTIPATION

1988 – 2004

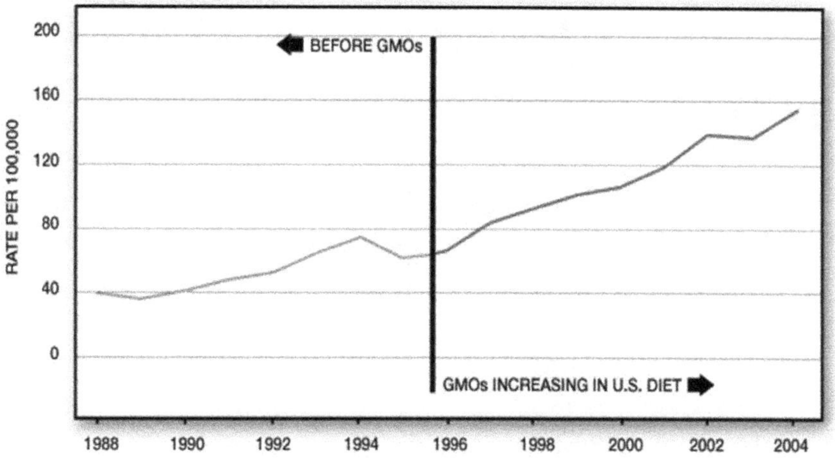

GASTROINTESTINAL INFECTIONS

1988 – 2004

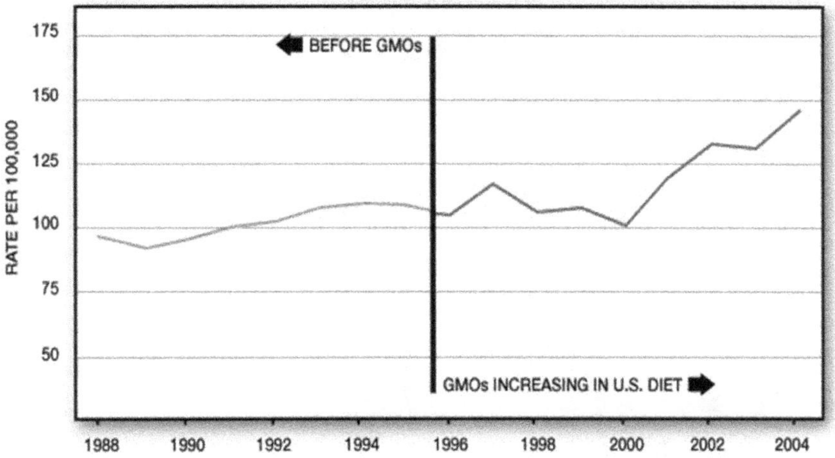

DIABETES IN AMERICA
1988 – 2011

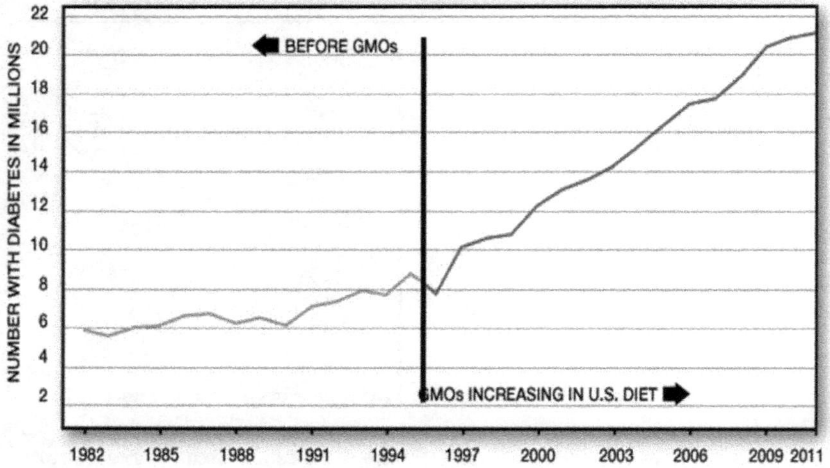

GASTROESOPHAGEAL REFLUX
1988 – 2004

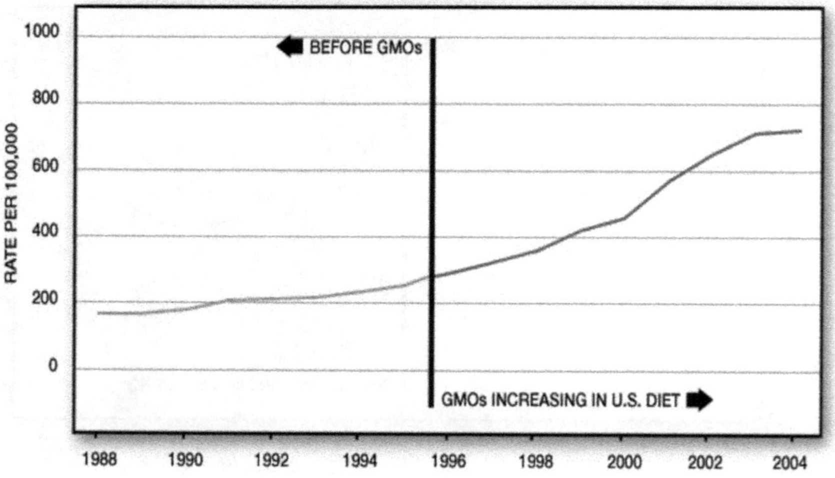

Effects of Genetically Modified Crops

The top ten genetically modified crops are corn, soy, cotton, canola oil from rapeseed, papaya, rice, tomatoes, and dairy products containing genetically modified bovine growth hormone rBGH (also rBST), potatoes, and peas. Today, in the US, 90% of corn, 91% of soybean, and 88% of cotton are genetically engineered. This leads to about 75% of processed foods. Genetic modification is not new. For hundreds of years, crops were altered through selective breeding. Genes were transferred during selective breeding. Most of the selective breeding was done by cross-pollination of the same family of crops. The new technology can select a particular gene characteristic, initially foreign to the host, and insert it into a plant or animal that has been selected to receive this characteristic. As an example, a gene from a fish living in very cold waters can be injected into a strawberry plant to be frost tolerant. The only problem that has not been answered is what beneficial plant trait it has replaced or what health impact this new breed of organism will have an effect.

Gene-modified (GM) ingredients appeared in over 65% of all US processed foods between 1997 and 1999. This food alteration was fueled by Supreme Court decision allowing a patent on life-forms. The first commercially grown genetically modified food crop was a tomato plant (the Flavr Savr). It was made resistant to rotting and released into the market in 1994 without any lengthy safety studies or labeling. Some of the reasons given for using GMO crops propagated by the biotech industry, sold to the public, were the following statements:

1) GMOs are needed to feed the world.
2) GMOs have been thoroughly tested and proven safe.
3) GMOs increase yield.
4) GMOs reduce the use of agricultural chemicals.
5) GMOs can be contained and therefore coexist with non-GMO crops.

It is interesting to note that all five of these statements were proven to be nothing but a well-organized propaganda so that the general public would accept the GMO crops. All the glowing words about helping the planet were just a front. All five are pure myths, blatant lies, regarding benefits of this new technology. The pro-GMO lobbyists claim that 1,700 studies show GMO foods to be safe. A closer analysis of many of these studies show risk to human health. The long-term environmental and human health impacts of genetically modified organisms have not been established. Since the early 1990s, genetically modified food crops have been introduced into the food market.

Let's take a look at the five items propagated by the biotech industry that are pushing these GMO food crops on the world.

1) GMOs are needed to feed the world: Reports are coming in that there are massive crop failures occurring throughout the world, resulting in farmers committing suicide. Crop failure pushed them into bankruptcy.
2) GMOs have been thoroughly tested and proven safe: No long-term studies were performed to prove the safety of GMO crops in food supply. On the contrary, the few long-term studies performed indicate high risks to human health. For example, the Seralini studies showing tumors on rats and affecting sperm production, indicating potential infertility. Also, reports are coming in regarding deadly effects on our pollinators—bees and the monarch butterflies, to name a few.
3) GMOs increase yield: There are numerous instances that the modified crop fails to materialize. There are also reports that the nutritional value is quite a bit inferior.
4) GMOs reduce the use of agricultural chemicals: Reports are now coming in that pests and weeds have become more resistive to these agricultural chemicals, requiring increase use amounts of same.
5) GMOs can be contained and therefore coexist with non-GMO crops: Latest reports indicates cross-pollination, damaging conventionally or organically grown crops. Farmers who have their crops contaminated are being sued by the biotech companies for patent infringement.

http://www.BlaylockWellness.com/GMO

http://www.wow.com/Effect-Of-Genetically-Modified

http://www.responsibletechnology.org/gmo-dangers/65-health-risks/Inotes-29k

http://www.ncbi.njm.nih.gov/pubmed/12746139

http://www.enhs.umn.edu./current/5103/gm/harmful.html-9k, reading

There is an organization that publishes "GMO Myths and Truths." They just published their second edition, released on May 16, 2014, that can be accessed by the following link—http://gmo-mythsandtruths.earthopensource.org.

I highly recommend reading what this organization has to say regarding GMO food. "GMO Myths and Truths" offers a one-stop resource for the public, campaigners, policymakers, and scientists opposing GMO industry's attempts to control our food supply." The authors of "GMO Myths and Truths" are not alone in doubting the safety of GMOs. In late 2013, nearly three hundred scientists and legal experts signed a statement affirming that there was "no scientific consensus on GMO safety." http://www.ensser.org/increasing-public-information/no-scientific-consensus-on-gmo-safety/

Additional information regarding these findings are on http://earthopensource.org/index.php/reports/gmo-myths-and-truths#st-hash.2rNvtUezM.dpuf

Top Ten GMO Food Crops

The top ten genetically modified food crops are corn, soy, cotton, canola oil from rapeseed, papaya, rice, tomatoes, dairy products containing genetically modified bovine growth hormone rBGH (also rBST), potatoes, and peas. Let's analyze some of these modified food crops as to what gene change these crops underwent.

Corn

The corn seed has been modified in such a way to create its own insecticide. A Bt gene from bacteria *Bacillus thuringiensis* is inserted into the corn crop. This enables the corn crop to produce an insecticide that is harmful to insects. When the insect eats any part of the plant, it then dies due to the insecticide contained in the plant. What is Bt toxin? *Bacillus thuringiensis* (Bt) is a bacteria that produces proteins which are toxic to insects. It is in the same family of bacteria as *B. anthracis*, which causes anthrax, and *B. cercus*, which causes food poisoning. The reason for its use in agriculture is that it is toxic to many insects that can disrupt farming operation; therefore, it is widely used as a biological pesticide. Bt can be sprayed or it can be added to the DNA of a genetically modified crop. The other reason for its application in agriculture is that it is cheap to produce.

The Bt corn is particularly used to control the European corn borer (ECB), which is considered to be most damaging to corn crops.

The other instance to corn crops is the damaging influence of weeds. The other genetic modification to corn crop was the addition of herbicide tolerant genes into the corn genome. Another genetic modification to corn seeds was to make them herbicide resistant. The predominantly used herbicide is Monsanto's Roundup, which contains glyphosate. This was inserted into the corn seed to make them herbicide resistant. The reasoning was that making some of the crops herbicide resistance, the farmers would use lesser amount of herbicides or pesticides in weed and/or insect control. The glyphosate becomes systemic throughout the plant from the genetically modified seed.

The gene that confers tolerance to glyphosate was discovered in a naturally occurring soil bacterium. Additional gene was added, which is glufosinate-ammonium tolerant, also derived from a naturally occurring soil bacterium. It was hoped that these herbicides, glyphosate, and glufosinate-ammonium provide a broad spectrum of weed control. The only consideration taken into account was for the farmers to minimize their insecticide/herbicide applications and diminish the damage to their crops—a very good reason and lofty goal for the agricultural industry. The only problem is that all these genetic modifications were not subjected to long-term health effect studies or the impact on our ecological system. Also, no nutritional evaluation analyses were done.

The following is an analysis done on Monsanto's GE corn regarding the nutritional value versus an organic grown corn. The report was given to Moms Across America by an employee of De Dell Seed Company. A picture of the nutritional differences between genetically engineered (GE) and non-GE corn clearly indicate that the GE corn is *not* equivalent to the organic version. This was one of the premises by which genetically engineered crops were approved in the first place. Another myth disproved. Below are the differences from that nutritional analysis:

Calcium:
GMO corn – 14 ppm
Non-GMO corn – **6,130 ppm** (437 times more)

Magnesium:
GMO corn – 2 ppm
Non-GMO corn – **113 ppm** (56 times more)

Manganese:
GMO corn – 2 ppm
Non-GMO corn – **14 ppm** (7 times more)

The GMO corn was found to contain **13 ppm of glyphosate**, an ingredient in the herbicide Roundup, zero in non-GMO corn. This is significant. The Environmental Protection Agency's (EPA) "safe" limit level for glyphosate in water supplies is **0.7 ppm**. Europe has an allowable limit of **0.2 ppm**. It has been reported that levels as low as 0.1 ppm have caused organ damage in animals. At 13 ppm, GMO corn has eighteen times more than the allowable safe level of glyphosate set by EPA. The other problem is that it cannot be washed off, since it is systemic to the plant and therefore ingested with our food.

http://www.articles.mercola.com/.../2013/04/30/monsanto-gmo-corn.aspx

Another study done by Profit Pro that tested GMO corn indicated that it contains a number of elements absent from traditional corn, namely chlorides, formaldehyde, and glyphosate. These elements do not appear naturally in corn, but they were present in GMO sample corn to the tune of 60 ppm of chloride, 200 ppm of formaldehyde, and 13 ppm of glyphosate. The disturbing information is that GMO corn contained high levels of formaldehyde. According to Dr. Huber, this level of formaldehyde is toxic to animals. It was also observed that the animals given a choice of GMO feed versus non-GMO feed will always choose the non-GMO feed. Mice fed with GM corn were discovered to have smaller offspring and fertility problems.

For more information, you can access the following links:

http://www.action.fooddemocracynow.org/sign/dr_huber_warning

http://www.gmowatch.org/latest-listing/51-2012/14164-glyphosate-and-gmos

http://www.articles.mercola.com/.../archive/2013/10/06/dr.huber-gmo-foods-.aspx

http://www.organicconsumers.org/articles/article_27367.cfm_before_its_news_.com.EU

One of the more important scientific studies was done by Professor Gilles-Eric Seralini in France on genetically modified NK603 corn and Roundup herbicide. The study looked at toxicity, but it observed tumors in rats as a byproduct of GMO exposure. This was the first time that a long-term study was done on either GMO crop and/or Roundup.

Critics were quick to discredit the Seralini study that it deviated from the norm to be accurate or it used the wrong rats. The problem is that there were no long-term studies done by anyone. The Sprague-Dawley rats are used in every GMO food study since they are a good crossover for assessing how humans will respond to a particular exposure. These rats are prone to tumors as are humans, as evidenced by the fact that almost all toxicity studies, even carcinogenicity studies, use the Sprague-Dewey rats as subjects. On top of that, **it's the results that count.**

http://ww.naturalnews.com/047162_Seralini_study_GM_corn.html

http://www.GMOEvidence.com

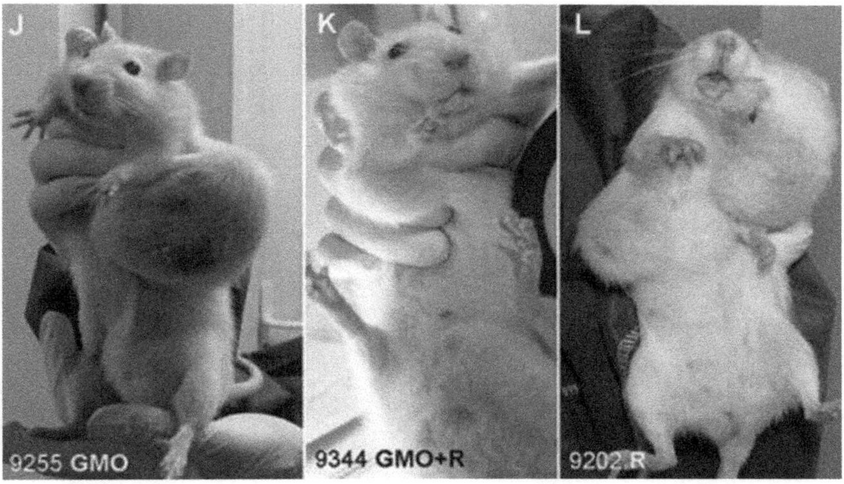

"The study, led by Gilles-Eric Seralini of the University of Caen, was **the first ever study to examine the long-term (lifetime) effects of eating GMOs**. You may find yourself thinking it is absolutely astonishing that no such studies were ever conducted before GM corn was approved for widespread use by the USDA and FDA, but such is the power of corporate lobbying and corporate greed.

The study was published in *The Food and Chemical Toxicology Journal* and was presented at a news conference in London.

Findings from the study

Here are some of the shocking findings from the study:

- Up to 50% of males and 70% of females suffered **premature death**.

- Rats that drank trace amounts of Roundup (at levels legally allowed in the water supply) had a **200% to 300% increase in large tumors.**
- Rats fed GM corn and traces of Roundup suffered **severe** organ damage, including liver damage and kidney damage.
- The study fed these rats **NK603**, the Monsanto variety of GM corn that's grown across North America and widely fed to animals and humans. This is the same corn that's in your corn-based breakfast cereal, corn tortillas and corn snack chips.

The *Daily Mail* is reporting on some of the reaction to the findings: "*France's Jose Bove, vice-chairman of the European Parliament's commission for agriculture and known as a fierce opponent of GM, called for an immediate suspension of all EU cultivation and import authorizations of GM crops. 'This study finally shows we are right and that it is urgent to quickly review all GMO evaluation processes,' he said in a statement. 'National and European food security agencies must carry out new studies financed by public funding to guarantee healthy food for European consumers*" (http://www.dailymail.co.uk/sciencetech/artic . . .).

Read the study abstract

The study is titled "A Comparison of the Effects of Three GM Corn Varieties on Mammalian Health." Read the abstract here: http://www.biolsci.org/v05p0706.htm

That abstract includes this text. Note: "hepatorenal toxicity" means toxic to the liver: "*Our analysis clearly reveals for the 3 GMOs new side effects linked with GM maize consumption, which were sex— and often dose-dependent. Effects were mostly associated with the kidney and liver, the dietary detoxifying organs, although different between the 3 GMOs. Other effects were also noticed in the heart, adrenal glands, spleen and haematopoietic system. We conclude that these data highlight signs of hepatorenal toxicity, possibly due to the new pesticides specific to each GM corn. In addition, unintended direct or indirect metabolic consequences of the genetic modification cannot be excluded.*"

Here are some quotes from the researchers:

"This research shows an extraordinary number of tumors developing earlier and more aggressively—particularly in female animals. I am shocked by the extreme negative health impacts" (Dr. Michael Antoniou, molecular biologist, *King's College London*).

"We can expect that the consumption of GM maize and the herbicide Roundup, impacts seriously on human health" (Dr. Antoniou).

"This is the first time that a long-term animal feeding trial has examined the impact of feeding GM corn or the herbicide Roundup, or a combination of both and the results are extremely serious. In the male rats, there was liver and kidney disorders, including tumors and even more worryingly, in the female rats, there were mammary tumors at a level which is extremely concerning; up to 80% of the female rats had mammary tumors by the end of the trial" (Patrick Holden, Director, Sustainable Food Trust).

There is a very interesting video, titled "What is a GMO" by Nutiva. Recommend viewing it. http://www.youtube.com/nutiva?feature=w . . .

Soy

Soy has been genetically modified to resist herbicides. Soy products that contain GM are soy flour, tofu, soy beverages, soybean oil, and other products that may include pastries, baked goods, and edible oil. Soy products are also in baby formulas and foods. Glyphosate tolerant soybeans contain high residue of glyphosate and its toxic breakdown byproduct AMPA (a-amino-3-hydroxy-5-methyl-4-isoxazolepropionic-acid). It mimics the neurotransmitter glutamate, excess amounts in brain are neurotoxic.

The biotech industry claims that GMO-soy is equivalent to conventional or organic soy. Yet they have never provided a concrete, independent study to illustrate that assertion. http://www.science.direct.com

Some of the more important nutritional differences reported are the following:

1) Glyphosate-tolerant GMO soybeans contain high residue of glyphosate and AMPA—a compound that mimics the effects of neurotransmitter glutamate.
2) Soybeans from different agricultural practices differ in nutritional quality.
3) Organic soybeans showed a significant healthy nutritional value.
4) The study rejects that GM soybeans as substantially equivalent as non-GM soybeans.

The soy samples for this study were grouped into three categories:

1) Genetically modified, glyphosate tolerant soy (GMO-soy)
2) Unmodified soy cultivated using conventional "chemical" regime
3) Unmodified soy cultivated using a conventional "organic" regime

The healthiest nutrient-rich soy was from the organic soybeans that had more sugars, more proteins and zinc, and less fiber than both conventional and GM soy. Organic soy also contained less total saturated fat and total omega-6 fatty acids than the other two soybeans. The GM soy contained residues of glyphosate and the breakdown product AMPA. The study used thirty-five different nutritional and elemental variables to characterize each soy batch. The study was able to discriminate between the three batches of soybeans, namely, GM soy, conventional, and organic soybeans without exception. The study demonstrated substantial nonequivalence in nutritional composition characteristic. This raises the following question—Why are these toxic products used in hundreds of different food products as well as in baby formulas? http://www.ncbi.nlm.nih.gov/pubmed/24491722

Cotton

Cotton has been designed to resist pesticide. It is considered food since its oil can be consumed. Its introduction in Chinese agriculture has produced a chemical that kills the cotton bollworm. This reduces the incidences of pests not only in the cotton crops but also in neighboring fields of soybean and corn. The introduction of Bt cotton in India had some very undesirable effect. Some of the GM seeds provided to the farmers had crop failure, which is a far cry from the promise of increased yields. Numerous Indian farmers committed suicide. The same is true in France, where incidence of farmer suicides has increased. The Bt cotton seeds are more expensive than traditional seeds and have to be purchased annually from the biotech industry. A crop failure usually wipes out the farmer financially. The other reason is that these seeds require more water and pesticide applications but have failed to produce increased crop yields or none at all. According to the World Health Organization, this suicide trend is on the increase worldwide. This is contrary to the farmers promise by the biotech industry regarding increased crop yields resulting in better financial profits.

One of the more interesting articles regarding Bt cotton seeds was published by the Seattle Organic Restaurants under the title "Indian Farmers Committing Suicide as a result of Monsanto's GM Crops."

http://www.seattleorganicrestaurants.com/vegan-whole-foods/Indian-farmers-committing-suicide-monsanto-gm

Rapeseed

Rapeseed oil has a fascinating history. Rapeseed oil is the most toxic of all food-oil plants. Rapeseed oil contains high amounts of erucic acid and glycosides, which adversely affect our health. Erucic acid is a substance associated with heart lesions in animal studies. To minimize these adverse health effects, through monitored cross breeding, which is a form of genetic modification, the amount of the erucic acid and glycosides was diminished to a point where it was renamed LEAR oil (low erucic acid rapeseed). Scientists had found a way to replace all of rapeseed's erucic acid with oleic acid, a beneficial fatty acid, commercially known as canola oil. Canola oil was introduced to the American consumers in 1986. http://www.snopes.com/medical/toxins.canola.asp

Animal studies on low erucic acid rapeseed oil were first performed when the oil was first developed. The LEAR study results were mixed. Rats genetically selected to be prone to heart lesions developed more lesions on the LEAR oil and flax oil than those on olive oil or sunflower oil. The researchers speculated that the omega-3 fatty acids (not erucic acid) in LEAR oil could be the culprit. Rats genetically selected to be resistant to heart lesions showed no significant difference between the four oils tested. Researchers at Canadian Institute for Food Science and Technology looked at twenty-three different experiments involving rats at different independent laboratory facilities. The conclusion was that saturated fats were more protective against heart lesions but that high levels of omega-3 fatty acids correlated with high levels of lesions. A study done with high levels of omega-3 fatty acids do not pose a problem with adequate amounts of saturated fats. That study also indicates that with adequate saturated fats it helps the body to convert the omega-3 fatty acids into the long chain versions of EPA and DHA.

Modern oil processing removes the oil by a combination of temperature mechanical pressing and solvent extraction, usually using hexane. Canola oil also undergoes caustic refining, bleaching, and degumming, which involves high temperatures or chemicals of questionable safety. Canola oil is high in omega-3 fatty acids but it easily becomes rancid and foul smelling. It is deodorized by turning the omega-3 fatty acids into trans-fatty acids. A large portion of canola oil has been hardened through the process of hydrogenation, which introduces levels of trans-fatty acids into the final product that can be as high as 40%.

A genetically modified rapeseed, herbicide tolerant, was introduced in Canada in 1995. It was the Roundup Ready canola. In 2009, 90% of the Canadian crop was herbicide resistant. In 2005, 87% of US crop was herbicide resistant. A study in 2010 conducted in North

Dakota found glyphosate and/or glufosinate resistance transgenes in 80% of wild rapeseed plants. In rapeseed plants genetic modification was done to make the plant's tolerance resistant to glyphosate and/or glufosinate. Regulation varies as to use of herbicide-resistant canola by individual countries.

Rapeseed oil is an acetyl cholinesterase inhibitor. This compound is crucial to transmitting signals from nerves to the muscles. Could this be another contributor to the increase in multiple sclerosis? Some of the other side effects reported are respiratory illnesses, vision loss, constipation, anemia, heart disease, irritability, and low birth weights in infants. It can also cause severe red blood cell clotting, and depletion of vitamin E. Of interest is that a number of people have reported allergic reactions to foods cooked with canola oil. The reported symptoms are dizziness, vomiting, nausea, skin irritations, and sneezing.

One of the more interesting reports and maybe significant side effects was reported by John Thomas, author of *Young Again: How to Reverse the Aging Process.* Quite a few people remember the mad cow disease epidemic that was rampart in Great Britain. Rape oil was widely used in the animal feed between 1986 and 1991. Pigs and sheep went blind, behaved insanely, and attacked other animals and people. Everyone thought that this was some kind of viral disease called scrapie in sheep and pigs and mad cow disease in cattle. The surprising thing is that when rape oil was eliminated from the animal feed, the symptoms disappeared.

http://www.naturalnews.com/031550_canola_oil_side_effects.html

http://www.rense,com/politics5/dare.htm

http://www.westonprice.org/health-topics/the-great-con==ola/-74k

http://www.naturalenews.com/02951_canola_oil_fraud.html-157l

http://www.ncbi.nlm.nih.gov/pubmed/6833663

Papaya

Papayas are also known as papaw or tree melons. They contain a high level of vitamins and some enzymes like papain and bromelain, which are beneficial to human health. The papaya ringspot virus is a major problem in papaya cultivation in Hawaii. Hawaiian papaya crops were pretty much wiped out by the ringspot virus. This papaya ringspot virus (PRV) can infect many varieties of papaya. The infected plants grow more slowly, become stunted, and produce bitter, spotted fruit. The virus is usually spread by contact with other trees or through contact by people, animals, and also aphids or sucking insects. Virus diseases are contagious to other papayas and certain other plants. Symptoms of infected trees are deformed leaves and areas of light green and dark green spots. The leaf stems or petioles will also show dark green streaks, and the fruit will get ring spots. The fruits take a long time to ripen, and eventually the trees die.

This prompted to have papayas genetically modified to be ringspot virus resistant. To accomplish this modification, a certain viral protein coat gene was transferred to the papaya genome. The first GM virus resistant papayas were commercially grown in Hawaii in 1999 and were highly successful. Since about 80% of Hawaii's papaya are transgenic (GM) the long-term health effects of these GM papayas has not been studied and will only be known after decades of consumption. With the success of the Hawaiian papaya, other papaya variants are being developed in different part of the world that are resistant to the viruses.

There are two kinds of papayas commonly grown. One is sweet and red or orange like flesh, and the other has yellow flesh. The large red-fleshed papaya often sold in the US are commonly grown in Mexico and Belize. The South American countries remain fairly safe for organically grown papayas.

One of the more interesting papayas comes from the Philippines. Researchers reported that by intergenic hybridization between *Carica* papaya and *Vasconcellea quercifolia* papaya, they had developed conventionally bred, nongenetically engineered papaya that are proving resistant to PRV (ringspot virus).

This is probably a GM success story presently, but it depends on long-term observation of GM papaya ingestion. Genetically modified papaya is sold only in US and Canada. The European nations do not import any of these GM papayas.

http://www.gmo.compass.org . . ./14.genetically_resistantodified_papaya_virus_resistance.html_61j,

http://www.etahr.hawaii.edu/oc/freepubs/pdf/CFS-4A.pdf

http://www.extento.hawaii.edu/kbase/crop/type/papring.htm-13k

Rice

Rice is the world's most important food staple consumed by more than 50% of the planet's population. The domestic variety originated in China. China has a multitude of rice varieties. An organization, Greenpeace uncovered in 2011 GE (genetically engineered) rice that had illegally contaminated the rice food supply in some Chinese provinces, even though it was still in seed test stage. The risk is high of GE rice contaminating the different wild rice varieties, also ecological dangers.

The rice from Southeast Asia was genetically modified to increase amounts of vitamin A and is known as the golden rice. The other modification was inserting a gene from *Bacillus thuringiensis*, a soil bacterium. The Bt toxin is normally inactive and is only produced in spores by the bacteria under stressful, life threatening conditions. Organic farmers have used Bt spores under serious pest infections, but Monsanto obtained a patent on the same gene and engineered it into plants so they can produce the preactivated (modified) Bt toxin to kill insects.

There are many problems with such GE or GM plants. The GE plants produce the toxins during the entire growing period. Insects are continually exposed to this toxin. Studies also indicate that the Bt toxin is also released into the soil. Rats fed GM rice for ninety days had a higher water intake. They showed differences in blood biochemistry, immune response, and gut bacteria. Organ weights of female rats fed GM rice were different from those fed non-GM rice.

The big, significant Bt GM-fed to rats, was a significant increase in coliform bacteria.

Another form of GE rice is made by Bayer. This modified version is to withstand high doses of the herbicide glufosinate that is sprayed on rice fields to control a wide range of weeds. Glufosinate is considered to be harmful to human health and is phased out in Europe.

http://wwwbiosafety.ru/domingo.pdf

http://www.greenpeace.org/ . . . /food-agriculture/genetic-engineering/ge-rice/93k

http://www.da3na8pro7vhmx.cloudfront.net/ . . . ewed stufies on gm food health risks.pdf

http://www.go.nature.com/YZjRyx

http://www.i-sis.org.uk/rice.php-58k

Tomatoes

Tomatoes have been modified for longer shelf life to prevent them from early rotting or degrading. In a test conducted to determine the safety of GM tomatoes, some animal subjects died within a few weeks after consuming GM tomatoes. The first genetically modified tomato was the FLAVR SAVR, to extend shelf life. This was the first commercially genetically modified food on the market. This transgenic tomato had deactivated gene, meaning that the plant was no longer able to produce polygalacturonase, an enzyme involved in fruit softening. This made the ripening process longer. Today there are no longer any GM tomatoes in the market. The Flavr Savr GE tomato failed to achieve commercial success and was withdrawn from the market in 1997.

That does not mean that the scientists are not working on some modification using the tomato as a good basic plant for GM modification. Some of the research being done is to have the tomatoes avoid stresses like frost, drought, and increased salinity, which are limiting factors to growth. Other genes from various species have been inserted into the tomato hoping to increase resistance to various environmental factors.

The insect toxin from the bacterium *Bacillus thuringiensis* has been inserted into a tomato plant. Field tests indicated resistance to the tobacco hookworm, tomato fruitworm, tomato pinworm, and tomato fruit borer. Tomatoes have been altered in attempts to improve their flavor and/or nutritional content. Another group has tried to increase the levels of isoflavones, known for its potential cancer preventive properties. Tomatoes along with potatoes, bananas, and other plants are being investigated as vehicles for delivering vaccines. Clinical trials have been performed on mice using tomatoes expressing antibodies or proteins that stimulate antibody production targeted to norovirus, hepatitis B, rabies, HIV, anthrax, and respiratory syncytial virus. Presently, the tomatoes in the market are non-GM, but this may change in the future since there are numerous genetic tests on the tomatoes.

http://blaylockwellness.com/

http://www.wow.com/Modified+Tomato

The other high volume food crops genetically modified are potatoes and peas. More information can be accessed via the following links:

http://www.soilassociation.org/LinkClick.aspx

http://www.leopold.iastate.edu/sites/default/files/events/Chapter16.pdf

http://www.disabled-world.com/fitness/gm-foods.php

http://www.sciencebasedmedicine.org/killer-tomatoes-and-poisonous-potatoes/-218k

http://www.ecologyandsociety.org/vol4/issl/art11/-22k

http://www.gmfreecymru.org/pivotal_papers/review.html

GMO Lies vs. Reality

10 facts you need to know about Seralini biotech study

http://www.naturalnews.com/047162_GMOs_Seralini_study_GM_corn.html#ixzz3cscuWV8Q

(NaturalNews)

Not long after its publishing, Professor Gilles-Eric Seralini's landmark study on genetically modified (GM) NK603 corn and Roundup herbicide received considerable undue criticism from the mainstream scientific community, which clearly didn't approve of its findings. But this doesn't negate the fact that Seralini's study exceeded the standard criteria for honest scientific inquiry, being the only study of its kind to look at the long-term effects of GMOs on mammals.

Here are 10 things you need to know about Seralini's study that validate its findings and put to shame the liars who claimed that it was a fraud.

1) Seralini study looked at toxicity, not cancer. One of the major criticisms levied against Seralini's work alleged that it was a badly designed cancer study. But it was actually a toxicity study that just so happened to observe cancer as a byproduct of GMO exposure. And based on the criteria of a toxicity study, Seralini's work was both well-designed and well-conducted.
2) No other studies have looked at long-term GMO toxicity. Some have tried to claim that Seralini's findings deviate too much from the norm to be accurate. But no other long-term toxicity studies on either GMOs or Roundup have ever been conducted, so of course his findings are going to stray from the status quo.
3) Everyone uses Sprague-Dawley (SD) rats. Claims that Seralini's study used the wrong variety of rat, known as Sprague-Dawley (SD), are also invalid. Nearly every GMO food study ever conducted, including the pitiful, 90-day feeding studies conducted by Monsanto, has used SD rats as subjects since they're a good crossover for assessing how humans will respond to a particular exposure.
4) SD rats and humans are almost equally prone to cancer. As far as cancer risk, SD rats are almost just as prone to tumors as humans are, as evidenced by the fact that almost all toxicity, and even carcinogenicity, studies use

SD rats as subjects. And just like people, SD rats have a higher risk of cancer the older they get, which makes them especially appropriate for long-term safety studies.

5) Seralini's sample size was appropriate by normal standards. Critics of Seralini's allegedly small sample size should take a look at all of Monsanto's toxicity studies—they're exactly the same! The only difference is that Seralini actually looked at how rats are affected by GMOs and Roundup over the long haul rather than just 90 days.

6) If Seralini study isn't valid, neither are Monsanto's studies. The irony in the scientific elite declaring Seralini's study invalid is that, by the same standards, so are Monsanto's GMO safety studies. Since Seralini's study criteria matched or exceeded everything Monsanto and others have done, at least in terms of protocol—as previously mentioned, Seralini's study is the only one that looked at GMOs long-term—then either his study is valid or theirs aren't. You can't have it both ways.

7) Seralini study invalidates Monsanto 'safety' studies. Since the Seralini study is, in fact, completely valid, this poses a major problem for the establishment. By looking at how rats react to Monsanto's GM corn and Roundup herbicide over the course of two years rather than just three months, Seralini has proven that Monsanto's own short-term safety studies are inherently flawed.

8) Regulators are wrong; GM corn and Roundup are highly toxic to mammals. Whenever an industry-backed safety study has even suggested possible toxicity from GMO exposure in the short term, regulators have been quick to dismiss such findings as "not biologically meaningful." But based on Seralini's findings, the questionable toxicity outcomes in Monsanto's own research actually affirm that GMOs can lead to organ damage, cancer and premature death in the long term.

9) No governments even require long-term safety studies. It's interesting that the type of research Seralini conducted is

not required by any single government anywhere in the world. The basis for every GMO approval thus far has only been limited, 90-day feeding studies that, on occasion, suggest toxicity, but never actually prove it. This is convenient for Big Biotech, which never has to face the fact that its products cause cancer and death, as demonstrated in Seralini's study.

10) Seralini isn't alone in discovering GMO toxicity. While many of his peers have refused to vet his work due to political pressures, Seralini is supported by a number of other independent researchers who have come to similar conclusions about the toxicity of GMOs. GMOEvidence.com outlines many of these, including studies on the toxicity of both Roundup and GMOs in piglets, dairy cows, bees, various aquatic animals and other organisms. To learn more, visit:

GMOEvidence.com. Sources:
http://www.gmoseralini.org
http://science.naturalnews.com
http://www.naturalnews.com/047162_GMOs_Seralini_study_GM_corn.html#ixzz3csdI4CRt

Dairy Products with rBST Hormone

A percentage of cows in the US are injected with the genetically modified bovine hormone rBGH, also known as rBST, or its commercial name Prolisec. This hormone forces the cows to increase milk production by about 11% to 16%. The dairy industry for the past two decades produces milk with the artificial hormone rBST. The cows are injected with this hormone every two weeks to increase milk production. The surprising fact is that the dairy industry does *not* have to disclose that it is using this hormone. The natural hormone is produced by the pituitary gland in the cattle. The synthetic hormone is produced by genetic modification of the *E. coli* bacteria.

The commercial production was initially produced by Monsanto, who led the way to get this hormone approved by FDA. There were numerous studies done by a multitude of organizations and universities trying to establish the safety of this hormonal use.

Recombinant bovine somatotropin (rBST) was first protein produced through the use of biotechnology. Other examples of proteins produced by recombinant technology include renin used in production of cheese, lactase used to produce lactose free milk, and insulin used to control human diabetes. In lactating cows, somatotropin is a major regulator of milk production. rBST supplemented cows produce more milk and utilize nutrients more effectively. http://www.ansei.cals.cornell.edu/. . . /shared/documents/Recombinant_Bovine_Somatotropin.pdf

Animal and human risks

A report from Vermont non-profit advocacy group revealed that rBGH injected cows, part of Monsanto's financial study at the University of Vermont, suffered serious health problems. There was an alarming rise in the number of deformed calves, and an increase in mastitis. Mastitis is a painful bacterial infection of the udder, causing inflammation, swelling, and pus. Also blood secretions into the milk. These findings were confirmed by Health Canada's 1988 report, concluding that the use of rBGH hormone increases the risk of mastitis by 25%, affects reproductive functions, increases the risk of clinical lameness by 50%, and shortens the lives of cows. To treat mastitis outbreaks, the dairy industry uses increased amounts of antibiotics. This contributes to the antibiotic resistant bacteria. The federal government does not contain a testing program for antibiotic residue in the milk.

Milk from rBGH-treated cows contain higher levels of IGF-1 (insulin growth factor-1). People naturally have IGF-1, but elevated levels in humans have been linked to colon and breast cancer. Some scientists have expressed concern over the possibility that higher levels of IGF-1 in milk can cause cancer, although no direct connection has been established.

http://www.sustainable.org/797/rbgh-113k

http://www.bibliotecapleyades.net/ciencia/cienca_industry-weapons.300.htm-47k

A research scientist from Cancer Prevention Coalition detailed the risks of rBGH milk. According to the scientists, the dangers should have been obvious years ago when it was first presented to Monsanto. Instead, the company chose to ignore it. (Same philosophy and procedure as in the aspartame approval process, except ignored by the FDA director.) The rBGH-treated cows induce a 70% to 100% increase in the levels of the hormone IGF-1. This hormone is not destroyed by pasteurization or digestion as initially reported since in the dairy product it is protected by the casein and the dairy's buffering effect. Some of this hormone is absorbed by the body. The IGF-1 stimulates the proliferation of cancer cells. Recent studies in humans indicated that increased levels of IGF-1 predict increased rates in breast cancer and prostate cancer. "A recent article (Cancer Research, 55:2463-2469, June 1995) from Renato Baserga's Laboratory in Philadelphia has shown clearly that IGF-1 is required for the establishment and maintenance of tumors. The mechanism for this is that IGF-1 protects the cells from apoptosis (programmed cell death). The IGF-1 accelerates tumor growth and appears to affect the aggressiveness of tumors. As the level of IGF-1 is decreased, cell death can take place. We are talking about IGF-1 levels of 10 nanograms per ml, that is 0.00001 milligram per ml."

"The widespread consumption of rBGH supplemented milk is therefore an experiment on an unsuspecting population that could have horrendous consequences and overwhelm the health care system . . . The risk to benefit the nation of this experiment is clearly NOT in favor of the consumer."

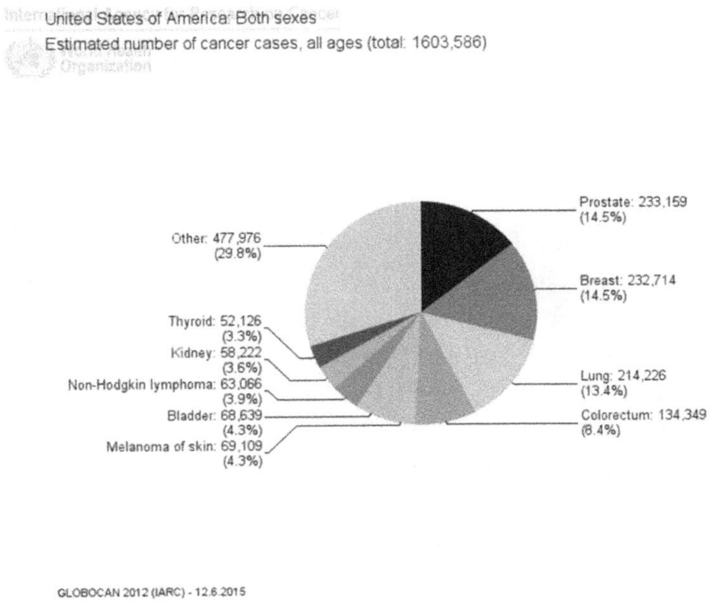

A study by Susan Hackinson, ScD, of Harvard Medical School, has showed that this hormone increases the risk of developing breast cancer sevenfold in women under fifty and that it increases a man's risk developing prostate cancer ninefold. Dr. Jenny Pompilio with Kaiser Permanente in Oregon says, "It's been known for years that this particular hormone is linked with cancer effects on the endocrine system."

Professor Samuel S. Epstein, MD, environmental medicine, University of Illinois, Chicago, School of Public Health, responded in a January 23, 1998, article in *Science*, "that men with high blood level of naturally occurring hormone insulin-like growth factor (IGF-1) are over four times more likely to develop full-blown prostate cancer then men with lower levels." He further noted "that the IGF-1 markedly stimulates the division and proliferation of normal and cancerous prostate cells and it blocks the programmed self-destruction."

These findings are highly relevant to any efforts to prevent prostate cancer, whose rates have escalated by 180% since 1950, which is now the commonest cancer in nonsmoking men with an estimated

185,000 new cases and 39,000 deaths in 1998. Now in 2014, this epidemic has not decreased but significantly increased.

Dairies that use this dangerous hormone don't even have to disclose that information to consumers. The FDA didn't stop at forbidding GM labeling; it went one step further. In 1994, they wrote a regulation requiring that any food label describing the product as bovine growth hormone free, must also include these words: **The EPA has determined no significant difference has been shown between milk derived from rBST and non-rBST treated cows. One of the biggest lies. Whose side is FDA on?**

The European Union even investigated the FDA's faulty approval process and concluded that the hormone's safety had never been proved. Canadian scientists also analyzed the FDA's approval process and wrote a lengthy and scathing report. One can read online under "Gaps Analysis Report." The report recounted omissions, contradictions, weaknesses, and gaps in the FDA's approval process. It concluded that **"the required long-term studies to ascertain safety were NOT conducted. Hence, the risk of sterility, infertility, birth defects, cancer and immunological disorder was not addressed."**

The agency continues to allow milk from cows treated with rBST. Unless it's labeled **"organic" or "rBST free,"** everything in the dairy section of the grocery store, including yogurt and cheese, probably includes milk from cows injected with this hormone.

The haunting questions is, why is FDA and/or EPA, supposedly the guardians of our health, allowing all this GMO food and herbicides, pesticides, and insecticides, allowed into our food supply? Why would FDA and EPA place the population at such astronomical health risks? With the help of FDA, GMO food has spread throughout our food supply. It turns out that the Deputy Commissioner for Policy at the FDA in 1993 had strong ties to Monsanto. He had worked for the company before joining FDA. In fact, after helping to secure the approval of rBST, he went back to Monsanto to work as its vice president for public policy. He's once again at the FDA, working closely with the Obama administration on food safety. Are we kidding each other? A person who was instrumental in getting a toxin like rBST approved is now at the helm of food safety or more likely

mass population poisoning. Over the past two decades, the FDA has been doing everything it can to keep us in the dark—to keep us from finding out how dangerous GM foods really are. "**We've all become unwitting guinea pigs in an experiment of massive proportions.**"

CAFOS

To make matters worse, the dairy farms use these so-called CAFOS (confined animal feeding operation) to feed the cattle in confined spaces and using GMO feed. As can be seen from the above picture, these animals are confined in tight quarters. They are not allowed to roam freely in pasture. No wonder that meat, pork, and poultry do not taste as good as from animals that are pasture raised, being able to roam freely and grass fed. A good comparison is eating an egg from a pasture chicken as opposed from a CAFO farm.

The question then comes to mind: Did the studies take this into consideration the confined space, the use of GMO feed, and the high usage of antibiotics? Most studies do not mention this wide range of suspected artificial chemical's effect on human health. How about water supply, did it contain fluoride? Most of the studies that I have browsed are of short duration. The most common statement issued is "no significant difference." What this implies that there is an effect, but due to the nature of the study length, it is classified as not significant to the researcher. The problem with this statement is that there is an effect, but due to the short time frame of the study, it is considered insignificant. On long-term studies, quite often these insignificant studies become significant as was demonstrated by the Seralini study on the long-term effect of GM corn in rats that came down with tumors.

Additional problems encountered by independent scientific organizations that want to do long-term studies, the biotech companies do not provide them with the necessary seeds to do that. Instead, these companies do their own studies, or farm these out to some universities with stipulation to come back with favorable results, or these academies lose their financial support. Then they have an inside person in authority at FDA or EPA to get a favorable approval ruling. This is done regardless of studies indicating adverse health effects.

A good background reading on the subject is Dr. Samuel S. Epstein's book titled *What's In Your Milk?* Additional good sources are:

1. American Institute for Cancer Research. Number of US Cancer Cases Expected to Rise 55 Percent by 2030
2. Medical News Today. US Obesity rates on the rise: 113 million by 2022
3. Elizabeth Weil. New York Times Puberty Before Age 10: A new "Normal?"
4. Food Safety—From The Farm To The Fork Report on Public HealthAspects of the Use of Bovine Somatotropin - 15-16 Naech 199
5. World Wire What's in Your Milk
6. New York Times Eli Lilly to Buy Monsanto's Dairy Cow Hormone for $300 Million
7. American Cancer Society Recombinant Bovine Growth Hormone
8. Hawsawi Y, El-Gendy R, Twelves C, Speirs V, Beattie J. Insulin-like growth factor - Oestradiol crosstalk and mammary gland tumourgenesis. Biochim Biophys Acta.

2013 Nov 2. pii:50304-419X(13)00049-4.doi:10.1016/j.bbcan.2013.10.005
9. Savvani A, Petraki C, Msaouel P, Diamanti E, Xoxakos I, Koutsilieris M. IGF-IEc expressin is associated with advanced clinica and pathological stage of prostate cancer. Anticancer Res, 2013 Jun; 33(6):2441-5
10. Kuklinski A, Kamocki Z, Cepowicz D, Gryko M, Czyewska J, Pawlak K, Kdra B. Relationship between insulin-like growth factor i and selected clinico-morphological parameters in clorectal cancer patients. Pol Przegl Chir. 2011 May;83(5):250-7.doi:10.2378/v10035-011-0039-z.
11. Wang Z, Wang Z, Liang Z, Liu J, Shi W, Bai P, Lin X, Magaye R, Zhao J. Expression and clinicalsignificance of IGF-1, IGFBP-3,and IGFBP-7 in serum and lung cancer tissues from patients with non-small cell lung cancer. Onco Targets Ther. 2013 Oct.16;6:1437-44.doi:10.2147/OTT.551997.
12. Geraci MJ, Cole M, Davis P. New onset diabetes associated with bovine growth hormone and testosterone abuse in a youngbody builder.
Hum ExpToxicol. 2011 Dec;30?(12):2007-12. doi:10.1177/0960327111408152. Epub 2011 May 9.

GMO Toxicity

The giant biotech Syngenta was charged for covering up livestock deaths

The giant biotech company Syngenta has been charged for covering up the death of livestock because of its genetically modified Bt corn. The story started back in Germany when the cows of a farmer suffered and suddenly died from a mysterious disease after they were exclusively fed on Syngenta's GM Bt 176. The farmer took Syngenta to court and Syngenta had to pay the farmer forty thousand Euros but the giant biotech completely denied that its GM corn caused the death of the livestock.

Syngenta dead GMO animals:

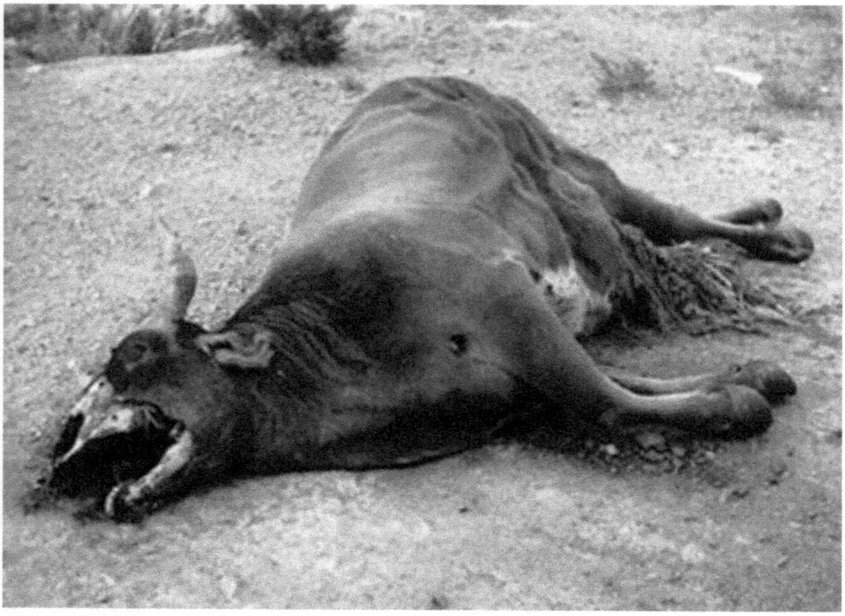

However, few years later, the farmer found out that in a feeding study, Syngenta's GM corn killed four cows in just two days. Now, few German farmers have come together and have taken Syngenta to the court for the death of famers' cattle and covering up the deaths of livestock in the study. Most importantly, Hans-Theo Jahmann, the head of Syngenta, is charged for not sharing his prior knowledge about the feeding study that resulted in the death of livestock in the first civil court. However, the German farmer isn't the only case that shows the toxicity of Bt corn.

According to Institute of Science and Society, more than 1,800 sheep were reported dead after grazing on Bt cotton crops from severe toxicity in four villages in Andhra Pradesh, India. Also in the last three years, ill workers and dead villagers have been reported in the same village.

Mass Deaths in Sheep Grazing on Bt Cotton

Also in 2003, in the Philippines, there have been several reports of death of livestock and fever, chest pain, respiratory, intestinal, and skin ailments of villagers who lived close by Bt maize fields.

Dozens Ill and Five Deaths in the Philippines and GM Ban Long Overdue

What's amazing is that in all the cases, Bt toxin has led to sudden death and mysterious illnesses in humans and livestock. In fact, in 2012 the Canadian study showed large percentages of both glyphosate and glufosinate (glyphosate is the active ingredient in Roundup used as herbicide for GM crops) in the blood of nonpregnant women. Also Bt toxin was found in the blood of 95% of pregnant women and 83% of their unborn fetuses. Recently, another long-term French study by Gilles-Eric Seralini also linked Bt corn to severe kidney and liver toxicity and giant tumors in rats.

Conflict of interest

Biotech is bad science shoved down our throats as science. It is impossible for Monsanto's scientists to claim no evidence of GMO harms, when other independent scientists have continuously reported severe toxicity and even death in laboratory animals. But Roundup is Monsanto's best-selling herbicide sold worldwide and it makes up for 43% of the company's income. Also more than 80% of the world's corn and soy have been genetically modified and are owned by the giant biotech companies including Monsanto. See more at: http://www.seattleorganicsrestaurants.com/vegan-whole-food/GMO-toxicity-bt-corn-syngenta.php#sthash_oLkupVXT.dpuf

Interesting to note that the Obama administration "quietly ended the Department of Justice's investigation of anticompetitive practices in the US seed market for the crops that cover the bulk of US farmland (corn, soy, and cotton), the seed trade is dominated by 5 companies: Monsanto, DuPont, DOW, Syngenta and Bayer.

Monsanto supplied nearly all genetically modified traits commonly used in those crops which it licensed to its rivals for sale in their seeds."

At the same time, many scientific community members have warned the authorities about toxicity of Roundup, among them Dr. Huber, who warned the secretary of agriculture about toxicity of Roundup. Dr. Huber also warned USDA about approving genetically modified alfalfa, which can contaminate other organic and non-GMO crops. But thanks to Obama's administration and his pro-GMO policies, Monsanto's people are assigned in key positions in both FDA and USDA.

In fact, Dr. Huber's letters about Roundup toxicity to Tom Vilsack (the secretary of agriculture who was awarded by biotech industries as the governor of the year) was ignored.

Also both USDA and FDA approved genetically modified alfalfa and Roundup knowing that it can cause birth defects, infertility, cancer, autism, gastrointestinal issues, and many other health problems. This is all happening when the American public is kept in dark because biotech companies spent $46 million against California GMO labeling.

India, in the beginning of 2013, joined sixty-four other countries, who have already required GMO labeling, to have GMO labeling. Also, twenty-seven other countries, including Zambia, Benin, and Serbia have banned GMOs imports and cultivation.

http://seattleorganicrestaurants.com/vegan-whole-foods/gmo-labeling-washington/

Resources:

http://occupymonsanto360.org/2013/03/02/syngenta-charged-for-covering-up-livestock-deaths-from-gm-corn/

http://www.gmwatch.org/latest-listing/1-news-items/13926-syngenta-charged-with-lying-over-cattle-deaths

GMO Health Dangers

Genetically modified foods—are they safe?

The American Academy of Environmental Medicine (AAEM) doesn't think so. The academy reported that "several animal studies indicate serious health risks associated with GM food," including infertility, immune problems, accelerated aging, faulty insulin regulation, and changes in major organs and the gastrointestinal system. The AAEM asked physicians to advise patients to avoid GM foods. Before the FDA decided to allow GMOs into food without labeling, FDA scientists had repeatedly warned that GM foods can create unpredictable, hard-to-detect side effects, including allergies, toxins, new diseases, and nutritional problems. They urged long-term safety studies, but were ignored.

Since then, findings include the following:

* Thousands of sheep, buffalo, and goats in India died after grazing on Bt cotton plants.
* Mice eating GM corn for the long term had fewer and smaller babies.
* More than half the babies of mother rats fed GM soy died within three weeks and were smaller.
* Testicle cells of mice and rats on GM soy changed significantly.
* By the third generation, most GM soy-fed hamsters lost the ability to have babies.
* Rodents fed GM corn and soy showed immune system responses and signs of toxicity.
* Cooked GM soy contains as much as seven times the amount of a known soy allergen.
* Soy allergies skyrocketed by 50% in the UK, soon after GM soy was introduced.
* The stomach lining of rats fed GM potatoes showed excessive cell growth, a condition that may lead to cancer.

* Studies showed organ lesions, altered liver and pancreas cells, changed enzyme levels, etc.

Unlike safety evaluations for drugs, there are no human clinical trials of GM foods. The only published human feeding experiment revealed that the genetic material inserted into GM soy transfers into bacteria living inside our intestines and continues to function. This means that long after we stop eating GM foods, we may still have their GM proteins produced continuously inside us. This could mean the following:

* If the antibiotic gene inserted into most GM crops were to transfer, it could create super diseases, resistant to antibiotics
* If the gene that creates Bt-toxin in GM corn were to transfer, it might turn our intestinal bacteria into living pesticide factories.

Although no studies have evaluated if antibiotic or Bt-toxin genes transfer, that is one of the key problems. The safety assessments are too superficial to even identify most of the potential dangers from GMOs. See our Health Risks brochure and State of the Science report for more details and citations. Recent health studies provide growing evidence of harm from GMOs:

GM Corn Damages Liver and Kidneys
Meat Raised on GM Feed is Different
Roundup Could Cause Birth Defects
Genetically Modified Soy Linked to Sterility

Chapter 12

Monosodium Glutamate MSG

What is MSG?

Monosodium glutamate is a salt of an amino acid—glutamic acid. The glutamic acid is found in proteins where it is connected to, or bound to, other amino acids in long chains. The two forms of glutamic acid found in nature are L-glutamic acid and D-glutamic acid. The glutamic acid found in protein is L-glutamic acid only. This form of glutamic acid, the L form, does not have adverse effects on the human body, specifically the brain. Man has eaten food in the form of protein for a long time. As part of protein digestion, it is broken down into its constituent amino acids, and one of them is glutamic acid, the L form. The protein is hydrolyzed (broken down) in the stomach and small intestines through the action of hydrochloric acid and digestive enzymes in the human body. The human body controls the amount of glutamic acid converted from proteins this way and gets rid of the excess as waste. The excess glutamic acid is not stored in the body.

In Oriental cultures, the glutamic acid was an extract of a seaweed, which was used to enhance the flavor of food for hundreds of years. It was in 1908 that it was finally identified as glutamic acid. It was initially manufactured in 1910 and designed for use as a food additive called monosodium glutamate. In 1956, the Japanese succeeded in producing glutamic acid by means of bacterial fermentation, and large-scale production commenced. The manufactured free glutamic acid (MSG) always contains the D-glutamic acid, pyroglutamic acid, and various other contaminants in addition to L-glutamic acid. Natural glutamic acid does not contain contaminants. Processed free glutamic acid (MSG) introduced into the body via food additive is not subject to the normal digestive processes and subject to the elimination process as that by ingested protein glutamic acid.

MSG toxicity

MSG is a brain neurotoxin that can largely be considered a slow poison. It is often disguised as a flavor additive. It is added to canned vegetables, soups, processed meats. It is generally recognized as safe by the FDA. MSG is an additive that can cause health concerns when consumed consistently. The buildup of it in the system is termed MSG toxicity.

"The first published report of an adverse reaction to MSG appeared in 1968 (Kwok, R. H. M., called the Chinese restaurant syndrome. Letter to the editor. *N. Engl. J* Med 278:796, 1968). The first evidence that MSG caused brain damage in the form of retinal degeneration was published in 1957 (Lucas, D. R., and Newhouse, J. P., The toxic effect of sodium-L-glutamate on the inner layers of the retina. *AMA Arch Opthalmol* 58: 193–201, 1057), and the first published report of brain lesions, obesity, and other disturbances with monosodium glutamate was published in 1969 (Olney, J. W., "Brain lesions, obesity, and other disturbances with monosodium glutamate," *Science*, 164:719–721, 1969)."

Multiple studies show that MSG produced by every one of manufactured methods can kill brain cells, cause neuroendocrine disorders, cause or exacerbate neurodegenerative disease, and cause adverse reactions in both animals and humans.

The Code of Federal Regulations and the Food and Drug Administration (FDA) distinguish between two classes of commercially manufactured glutamic acid when glutamic acid is to be used as a food additive (Code of Federal Regulations Food and Drugs 21:

Parts 100–169, 1990; Code of Federal Regulations Food and Drugs 21: Parts 170–199, 1989).

> Class I. When glutamic acid is refined to approximately 99% glutamic acid, the FDA requires that the ingredient/product containing the 99% pure glutamic acid be identified on food labels as "monosodium glutamate.
>
> Class II. When protein is broken down into its constituent amino acids, and refinement results in an ingredient/product that is less than 99% pure glutamic acid, the product is referred to as a "hydrolyzed protein product" (HPP). This is typically stated on labels as "Natural Flavors." The manufacturer can say "There is no MSG in our product" as long as the MSG is less than 99% of the ingredient listed.

It has been demonstrated by Olney and others that HPP (hydrolyzed protein product) such as glutamic acid type cause hypothalamic lesions and neuroendocrine disorders. The FDA regulations require that products that contain MSG in its HPP forms must be labeled with their individual unique common or usual names. *However, the FDA does not require, and has refused to require, that the MSG in products that contain any source of MSG be identified.* The FDA even goes further in allowing MSG to be hidden in food. Nothing on the label of a product containing MSG reveals that the product contains MSG.

US food regulators and chemical companies are intentionally making it difficult to recognize it if a given food product contains MSG. Common names listed to hide MSG in the ingredients list are ***gelatin, calcium caseinate, hydrolyzed vegetable protein (HVP), textured protein, monosodium glutamate, hydrolyzed plant protein (HPP), yeast extract, sodium glutamate, glutamate, autolyzed plant protein, yeast food, yeast nutrient, glutamic acid, sodium***

caseinate, autolyzed yeast, vegatable protein extract, soy protein, hydrolyzed corn gluten, natural flavor, artificial flavor, and spice.

Ingredients listed in food and beverages that might contain MSG or that were processed with MSG are ***bouillon, broth, carrageenan, enzyme-modified food, fermented foods and beverages, beer, bourbon/scotch/whiskey/brandy/wine, other alcoholic drinks, flavors and flavorings, natural chicken flavoring, natural flavorings, natural beef flavoring, natural pork flavoring, smoke flavoring, barley malt, malt extract, malt flavoring, maltodextrin, pectin, protein-fortified foods, textured protein, seasonings, soy extract, soy protein, soy protein concentrate, soy protein isolate, soy sauce, stock, ultrapasteurized foods, whey protein, whey protein concentrate, whey protein isolate.***

http://www.kisswebpage.com/msg

The FDA has intentionally left natural flavor and spice without meaningful definitions, the chemical companies freely use these names to hide the presence of toxins in the food. The FDA allows MSG to be hidden in food. Many MSG ingredients are added to "flavor," "natural flavors," "stock," or "broth." Other names are Accent, Ajinomoto, and natural meat tenderizer. Even the common or usual names of those containing MSG ingredients need not be disclosed. The glutamate industry goes to extreme lengths to ensure that consumers believe that the free glutamic acid is identical to the glutamic acid in intact protein.

http://www.earthclinic.com/cures/MSG.html

Documented Effects of MSG Consumption

The following is a list of reported monosodium adverse effects: epilepsy, vision disturbances, panic attacks, heart attacks, Parkinson's disease, Huntington's disease, ALS (amyotrophic lateral selerosis), Alzheimer's disease, brain lesions, retina damage, food cravings, depleted nutrients, hyperinsulinemia, stunted growth, crosses into the fetus, ocular (eye) destruction, liver damage, seizures, sciatica, diabetes, kidney damage, vastly increased chances of add, ADHD,

Asperger's or autism, severe headache, shortness of breath, chest pains, asthma, slowed speech, gastrointestinal disturbances, swelling, numbness of hands, feet, or jaw, chronic bronchitis, allergy reactions, irregular heartbeat, unstable blood pressure, pain in joints or bones, abrupt mood changes, tingling in face or chest, pressure behind eyes, difficulty swallowing, anxiety attacks, explosive rages, balance problems, dizziness or seizures, mini-strokes, fibromyalgia, MS (multiple sclerosis), tenderness in localized areas, neck, back, etc., chronic post nasal drip, sleep disorders, blurred vision, chronic fatigue, extreme thirst or dry mouth, difficulty concentrating and poor memory, arrhythmia, palpitations, diarrhea, stomach cramps, swelling of hemorrhoids, rectal bleeding, bloating, irritable bowel, joint pain, tendonitis, stiffness, disorientation, mental fatigue, lethargy, chills and shakes, vomiting, muscle aches (legs, back, shoulders, neck), depression, hoarseness, sore throat, balance problems, lethargy, cartilage connective tissue damage, gout like condition, gall bladder or gall bladder like problems, hematuria syndrome, sleepiness, asthma, nocturia, vaginal spotting, hives (internal and/or external), tongue swelling, and bags under eyes.

Information was from http://www.truthinlabeling.org/Recog.htm.

As can be deduced from these numerous symptoms, it is very difficult to diagnose if an adverse effect is due to MSG sensitivity or to potentially other food additives and/or toxins. MSG is also added to food but also to pharmaceuticals, cosmetics, and dietary supplements. What is even worse, the glutamate industry has convinced the medical community regarding the fictitious belief that reactions to MSG are allergic. That is *not* true. The adverse reactions to MSG are due to a toxin, the free glutamic acid. The only way to determine if a given reaction is to MSG is to ask the person regarding his food intake, cosmetic use, dietary supplement use, and/or medicine use.

MSG-sensitive people report reactions from simple skin rash to a severe depression or other life-threatening physical conditions. The severity of any adverse effect due to MSG depends on amount of free glutamic acid ingested. One should also remember that none of the symptoms of MSG toxicity are caused exclusively by MSG. Most, if not all, can be caused by various conditions as well as by other food additives. Difficulty in diagnosing MSG toxicity is further compounded by the industry practice of illegally advertising "No MSG Added" or "No Added MSG" if the level of MSG is less than 99%.

A physician and patient must work together to be able to identify, or rule out, MSG as a reactive trigger through analysis of a patient's food diary, restricting intake to totally unprocessed food and drink for about three weeks, then introducing items one at a time to help identify offending source(s) of MSG. To compound the problem, quite a few of the above listed adverse effects could potentially also be to aspartame, the artificial sweetener. According to Doctor Russell Blaylock, a neurosurgeon, author of *Excitotoxins: The Taste That Kills*, "excitotoxins literally excite brain cells to death." He has found that in addition to obesity, several neurological disorders come from long-term aspartame and MSG consumption. Dr. Blaylock explains that amyotrophic lateral sclerosis (ALS), also known as Lou Gehrig's disease, "is a neurodegenerative disease that primarily affects the anterior horn cells of the spinal cord. These are the primary neurons in the spinal cord that control muscle movements of the body." This constant increase of MSG as free glutamic acid in our food chain has corresponded to the surge in diabetes type 2, obesity, Alzheimer's disease, Parkinson's disease, and other neurological disorders. MSG is used liberally in all fast food places and in some sit-down restaurants. All snack foods and most packaged foods contain at least one form of disguised free glutamate, MSG, or under another name.

In 1997, MSG was also used on plants as a so-called growth enhancer under the name of AuxiGro. It was applied to the soil or sprayed on growing crops. The use of it has been discontinued in the United States but is used in Europe, Asia, Central and South America, New Zealand, Australia, and Africa. There are a number of other MSG containing agricultural products that are used as fertilizers or are being sprayed on the crops without restriction. Hydrolyzed fish protein and hydrolyzed chicken feathers are two of them. Additional reading regarding MSG

http://www.resourcesforlife.com/docs/item1225...

http://www.livestrong.com/sideeffectsofmonosodiumglutamate.html

http://www.meltdata.com/articles/harmful-effects-of-msg-html
http://www.evidenceofmsgtoxicity.blogspot.com/2011/11/cancermsg...

Additional usage of monosodium glutamate is in vaccines and 2-phenoxy-ethanol which are used as stabilizers in a few vaccines to help the vaccine remain unchanged when the vaccine is exposed to heat, light, acidity, or humidity.

http://www.cdc.gov/vaccines/vac-gen/additives.html

MSG as Crop Spray

In 1997 EPA approved a new crop spray called AuxiGro, which contains 29.2% pharmaceutical grade glutamic acid (MSG), for use on crops in the US. In 1995 FDA proposed that free glutamic acid be labeled due to its potentially deadly effect on individuals with asthma. The EPA at that time also noted that the "agency is not requiring information on the endocrine effects on these compounds at this time." It also stated that "waivers have been requested for acute toxicity, genotoxicity, reproductive and developmental toxicity, subchronic toxicity, chronic toxicity, and acute toxicity to non-target species based on GABA's ubiquity in nature, use as a pharmaceutical agent, favorable toxicological profile in chronic and other toxicology studies, and inconsequential exposure resulting from label-directed uses." These waivers were granted.

The reason it was initially granted because it was used as a growth enhancer. The free glutamic acid is converted to gamma amino butyric acid (GABA) in vegetables much as it is in the human body. GABA in humans prompts the pituitary to release growth hormone. It also increases growth in plants.

The free glutamic acid, as L-glutamic acid found in protein, is beneficial. The manufactured L-glutamic acid contains D-glutamic acid, pyroglutamic acid, and other chemicals referred to as contaminants. It is manufactured in chemical plants by certain selected genetically engineered bacteria, feeding on a liquid nutrient medium, excreting the free glutamic acid they synthesize outside of their cell membrane into the liquid medium. The free glutamic acid found in protein and the free glutamic acid involved in normal human body function are unprocessed and contain no contaminants. In protein, amino acids are found in balanced combinations. Use of free glutamic acid as a spray on crops throws the amino acid balance out of balance.

Some of the things that have to be considered are the following:

1) No one knows what the long-term effects of spraying processed free glutamic acid on crops will be.
2) Free glutamic acid will be absorbed into the body of the plant.
3) There is some residue left on the plants. No studies of either the amount of least amount of processed free glutamic acid needed to cause a reaction in an MSG sensitive person.
4) The free glutamic acid is toxic to the nervous system.
5) Children are most at risk from the effects of processed free glutamic acid. Their underdeveloped blood-brain barrier leaves them at high risk. It has been demonstrated that infant animals fed processed free glutamic acid when young develop neuroendocrine problems, such as gross obesity, stunted growth, and reproductive disorders, later in life. They also develop learning disabilities.
6) No one knows the amount of glutamic acid to kill a single brain cell.
7) Free glutamic acid is a neurotransmitter. It causes nerves to fire nerve impulses throughout the nervous system.
8) Free glutamic acid is a neurotoxin. Under certain circumstances, it will cause nerves to fire repeatedly until they die.
9) Processed free glutamic acid ingested by laboratory animals in the form of monosodium glutamate in food, caused brain lesions and neuroendocrine problems.

The National Institute of Health recognizes glutamic acid as being associated with addiction, stroke, epilepsy, degenerative disorders (such as Alzheimer's disease, Parkinson's disease, ALS), brain trauma, neuropathic pain, schizophrenia, anxiety, and depression. The EPA's statement regarding the safety of AuxiGro sidesteps the issue of endocrine safety, the crux of MSG and GABA toxicity. MSG—specifically the free glutamic acid in MSG—was proven to damage brain cells in the hypothalamus, the part of the brain that directs the pituitary and the entire endocrine system.

Many patients with diabetes have immune systems that directly attack the enzyme they use to handle excess glutamic acid. Any benefit diabetics may get from eating pure vegetables now may be negated by eating the very thing sprayed on those vegetables that may have caused their disease. One in four develop a pituitary tumor in their lifetime. For more information on the pituitary tumors and the effects of excessive growth hormone, see http://www.pituitary.org.

The AuxiGro was given the trademark number of 75262576. The current status of this trademark is "CANCELLED – SECTION 8" as of September 16, 2011 in the US The problem still exists as a food additive in the form of free glutamic acid. It is still used as a spray worldwide, as listed in these areas: Europe, Asia, Central and South America, New Zealand, Australia, and Africa.

http://www.truthinlabeling.org/msgsprayed.html

One of the more interesting sidelines that surfaced from the use of AuxiGro as a fertilizer is that it was responsible for colony collapse disorder, which devastated beehives in the US The theory proposed was that bees use pheromones – chemical signals affecting nearly all their most important behaviors. The one suspected is a particular pheromone released by the queen that repels workers when she feels threatened. The dairy casein was chosen to make AuxiGro in spraying from airplanes that blanketed the countryside with the pheromone that tells the bees to abandon the queen. Once the spraying was stopped, new reports indicate that beehives are on the increase. Could that be a coincidence?

http:www.msgtruth.org/cropspr.htm

Additional information regarding MSG are in the following books:

Excitotoxins: The Ultimate Brainslayer, by James South MA,
Excitotoxins: The Taste That Kills, by Russell Blaylock, MD,
MSG: A Dangerous Excitotoxin

http://www.vitalearth.org
TruthInLabeling.org
http://www.naturalnews.com/026157
MSG food organic.html#ixzz3IQaFJnKA

To compound the problem, the biotech and agricultural conglomerate has introduced a new kid on the block—Sweetmix, made by a biotech firm called Senomyx.

Some scientific studies on monosodium glutamate:

Effect of systemic monosodium glutamate (MSG) on headache and pericardial muscle sensitivity.

Monosodium glutamate (MSG) a villain and promoter of liver inflammation and dysplasia.

Association of monosodium glutamate intake with overweight in Chinese adults the INTERMAP study.

Neuroprotective of extract of ginger (Zingiber officinale) root in monosodium glutamate induced toxicity in different areas male albino rats.

Senomyx

This is a high-tech research and development company using cutting-edge biotechnology and most recent genetic engineering in their search for food additive taste buds' enhancement. Senomyx uses its numerous workforce to develop patented flavor enhancers by using proprietary taste receptors based assay systems. A press release from watchdog group, Children of God for Life, discloses the fact that the receptors are made from HEK293. HEK293 stands for human embryonic kidney cells. The cells came from a healthy, electively aborted baby. (http://www.cogforlife.org/senomyxalert.htm)

These cells were harvested and used by a Dutch scientist in the 1970s. The cells were cloned, grew them in cultures and later sold for research. The offspring of these cells became widely used in the medical biotechnology industry to produce therapeutic proteins and viruses. Now these cells are used by Senomyx in their quest to identify its new tastes.

These proprietary assay taste receptor-based system provides scientists with biochemical responses and electronic readout when a flavor ingredient interacts with their patented receptor. This lets the researchers know whether or not their flavor enhancer is effective. According to Weston A. Price Foundation, "Senomyx's salt taste, savory flavor, and sweet flavors as well as all their other flavor enhancers – are purposefully so that they stimulate your taste buds without them actually tasting anything . . . Much like MSG, these flavor enhancers operate on the neurological level to produce these reactions . . . Since they are not actually ingredients but rather 'enhancers' they are not required to be listed in a package's ingredients except as 'artificial flavors' . . . and because very small amounts of the additive are used . . . Senomyx's chemicals have not undergone the FDA's usual safety approval process for food additives."

Biotech firm Senomyx is at the forefront of the flavor enhancer market, as they state on their website http://www.Senomyx.flavorenhancers.com: "Using isolated human taste receptors, we created proprietary taste receptors – based assay system that provide a biochemical or electronic readout when a flavor ingredient interacts with the receptor. To enable faster discovery of new flavors, we integrated our assays into a robot-controlled automated system that uses plates containing an array of individual fluid wells, each of which can screen a different sample from our libraries of approximately 800,000 artificial and natural candidate ingredients isolated from plants and other sources."

"Our high-throughput discovery and development process allows us to conduct millions of analyses of new potential flavor ingredients annually. This efficiency is impossible to achieve using commercial flavor discovery methods. As a result, we have identified hundreds of unique potential new flavor ingredients that could not be discovered using taste tests alone."

As the *New York Times* wrote, "Unlike artificial sweeteners, Senomyx's chemical compounds will not be listed separately on ingredients labels. Instead, they will be lumped into a broad category . . . 'artificial flavors' . . . already found on most packaged food labels."

Senomyx's salt taste, savory flavor, and sweet flavors—as well as all the other flavor enhancers—are purposefully developed to stimulate one's taste buds without them actually tasting anything. This fools the brain into thinking that one has tasted an intensely sweet or savory flavor. Senomyx's MSG-enhancer gained the "Generally Recognized as Safe (GRAS)" status from the Flavor and Extract Manufacturers Association (FEMA). FEMA is an industry-funded organization. The only safety testing was done on rats for about three months, and Senomyx maintains that their product is safe to eat since they are used in very low concentration (1 ppm).

Many food companies have already partnered with Senomyx. Among these are as follows:

* Ajinomoto Group—using the savory flavors in China for many years. They are also the world leader in food transglutaminase, a substance that alters proteins to allow unrelated meats to be seamlessly "glued" together, or to improve the general texture of any protein-rich food. They are the world's largest users of MSG.
* Cadbury/Kraft is the parent company with too many brand families to list separately. They are looking for ways to use several flavors, including sweet and cooling, in their candies and confectionaries. Outside sources say that Kraft may be planning to use Senomyx's sweet flavoring to reduce the sugar in powdered beverages like Kool-Aid.
* Campbell's Soup is still listed on the Senomyx website, yet recent news reports claim Campbell's has since disaffiliated when the two companies were unable to come to an agreement.

* Firmenich (a Swiss perfume and flavoring company) is working with Senomyx to use a sweet enhancer in products that are currently sweetened with sucrose, fructose, and stevia. They are currently building a market interest in North American sales using S2383. (S2383 is a sweetener enhancer produced by Senomyx.) A search on the Firmenich website yielded this quote: "Extracted from nature or imagined by our scientists, our ingredients bring new emotions, tastes, and originality to every creation."
* Solac makes soy-based foods including protein bars and infant formulas and stands to benefit from the bitter blocker program.
* Nestle, the largest worldwide food processing company, has reformulated many of their old products and also developed new ones using the savory flavors S336 and S807. They market coffee drinks and coffee creamers that utilize Senomyx technology enhancers. They are also pushing the European Food Safety Authority to grant approval to use these ingredients in the European Union, creating a new market for themselves.
* Pepsi Co. (which also includes the Frito-Lay, Tropicana, Quaker, and Gatorade brands) recently signed a four-year contract with Senomyx that included a thirty-million-dollar upfront payment from Pepsi to Senomyx to use their sweet enhancers.

References:

http:www.senomyx.com

http://www.nytimes.com/2005/04/06/business/06senomyx.html?pagewanted=[&_r=]

http://seekimgalpha.com/article//64944-senomyx-inc-q4-2007-earnings-call-transcript

http://www.signonsandiago.com/2011/mar/30/biotechs-use-drug-discovery-
tools-find-new-flavors/

http://www.businesswire.com/news/home/20110429005199/en/SENOMYX
-ANNOUNCESCORPORATE-UPDATE-QUARTER-2011-FINANCIAL

http://seekingalpha.com/aeticle/266729-senomyx-ceodiscusses-q1-2011-
Results-earnings-call-transcript?source=thestreetconference

http://seekingalpha.com/article/259259-a-taste-of-theflavor-company-sector

http://www.wikinvest.com/stock/Senomyx_(SNMX)/Salt_Enhancer_Progrm

http:articles.mercola.com/sites/artcles/archive/2011/05/04/has-your-meat-been-glued-together-why-you-need-to-avoid-this-dangerousprocess.aspx

The above articles appeared in *Wise Traditions in Food, Farming and the Healing Arts,* the quarterly journal of the Weston A. Price Foundation, in the Summer of 2011.

Chapter 13

Pesticides

Pesticide is any substance or mixture of substances meant for attracting, destroying, preventing, repelling, or mitigating any pest. The term pesticide applies to a number of different pest specific substances. Target pests include insects, plants (like weeds), pathogens, mollusks, and nematodes, to name a few. Following is a general list of types of pesticides:

1) Organophosphate pesticides: These affect the nervous system, most are insecticides. These have the same adverse effect on humans.
2) Carbamates: These also affect the nervous system, specifically disrupting an enzyme that regulates acetylcholine, a neurotransmitter.
3) Organochlorine insecticides: Examples include DDT, Chlordane, Toxaphene—these have been discontinued since they had an adverse health effect and the environment.
4) Pyrethrid pesticides: This was a synthetic pesticide developed along the line of naturally occurring pyrethrin, found in chrysanthemum. Some are toxic to the nervous system.
5) Sulfonylurea herbicides: There are numerous types of these herbicides specific for weed control.

6) Biopesticides: These are pesticides derived from natural materials as animals, plants, bacteria, and certain minerals.
7) Microbial pesticides: These consist of a microorganism as an active ingredient. Pesticide is relatively specific for its target pest. The most widely used microbial pesticides are subspecies and strains of *Bacillus thuringiensis*.

History

The first recorded use of a substance as a pesticide was in Mesopotamia, about 4,500 years ago, which was elemental sulfur dust. They also used poisonous plants for pest control. During the fifteenth century, people started using toxic chemicals like arsenic, mercury, and lead. In the seventeenth century a compound named nicotine sulfate was extracted from the tobacco plant and used as an insecticide. During the nineteenth century, pyrethrum and rotenone came into use. Pyrethrum was derived from the chrysanthemum and rotenone from roots of tropical vegetables.

Until 1950s, arsenic derived pesticides were dominant. With the discovery of DDT, the organochlorine pesticides became the predominant pesticide used. In 1970s the EPA was established and given the authority to regulate pesticide use.

In the late 1960s, it was discovered that DDT had a very adverse health effect on fish-eating birds, namely being able to reproduce. This was a serious effect on biodiversity. Rachel Carson's book *Silent Spring* magnified the biological toxic effect of DDT. DDT was banned from further usage in the US.

As a reminder of this toxic effect on our environment, please go to the following links:

http://www.prezi.com/fwgtzfptl1_q/environmental-pollution-of-stlouis-mi

http://www.stlouis.com/1/stlouis/water_andwastewater.asp

Environmental Effects

The use of pesticides raises a number of environmental concerns. A very large percentage, about 98%, of the sprayed pesticides and herbicides, reach a destination other than the target areas. This includes air, water, and soil. The pesticide or herbicide drift due to wind condition carry the suspended pesticide particles to other nontargeted areas. They contaminate our water system by run-off in rainstorms. They also contaminate our soil. This results in biodiversity reduction, has a lethal effect on our beneficial pollinators like bees and the monarch butterfly, and threatens numerous endangered species.

The other problem is that pests can mutate and become pesticide resistant. This necessitates a new pesticide. To counteract this, larger doses are used which contributes to worse pesticide pollution problem. Another bad effect of pesticides is that the killed insects due to the pesticide can be eaten by birds, which then can have adverse health effects. The chlorinated hydrocarbon pesticides usually dissolve in fats which retain them. Pesticides that evaporate into the atmosphere at relatively high temperature are carried considerable distances by the wind to an area of colder temperatures, condense, and fall to the ground with rain or snow. Adsorption of pesticide by the soil reduces bioavailability due to microbial degraders.

Herbicides

Herbicides are a family of chemicals that are used to kill plants, specifically weeds. One of the desirable characteristics is that these compounds should be biodegradable. *Biodegradable*, according to *Webster's Dictionary*, is that the product is "capable of being decomposed by bacteria or other living organism." Synonyms for the word include *ecological*, *perishable*, *environmental*, *green*, and *recyclable*. The meaning indicates that it is nontoxic, "a product that can be safely used in nature without side effects."

Herbicides used to kill weeds do not break down in the soil. Herbicides can be divided into two major categories selective and nonselective. The first group is used to kill one type of plant without harming another. The latter group is used to sweep a field clean of whatever is growing there. Herbicides are further divided into groups according to the path they take to kill the plant. Herbicides can be toxic to humans, depending on the amount of dosage. Usually a herbicide is applied at the rate of one pound per acre, thus it is at low concentrations. Potentially, wildlife can be damaged, like birds that come in contact during spraying. The runoff from treated fields can be injurious to a range of water organisms.

Herbicides can remain in the soil for up to thirty months, but most of the commonly used chemicals are broken down within eight weeks after application. The danger lies if the treated crop is harvested prior to that date of application and send to the markets. Ecological effect concern is that they may kill microorganism in the soil. The bacteria and fungi that decompose organic matter that make the soil fertile. The other concern is if the weeds may develop new strains that are resistant to the herbicides.

True biodegradability of herbicides is just a myth. The other myth is that some accepted products are used safely around humans and animals and are found in ground water and drinking water. This indicates that they have not been decomposed by bacteria or other living organisms. Although this is a minimal exposure and may be classified as safe, the question remains as to potential accumulation in the soil and/or ground water. The biodegradability promise presented by numerous commercial herbicide manufacturers, particularly who promote a product containing glyphosate, is widely claimed as being safe to humans and animals. A study was conducted by Cornell University feeding dogs food containing glyphosate, for up to two years without any side effects. But a study by Organic Consumers Association reported, "The product is in fact highly toxic to the soil life and significantly reduces the activity of nitrogen fixing bacteria." This is contrary to the report that states it is harmless and safe and breaks down leaving no residue.

A major change is occurring in the agricultural industry regarding farming of their land. The old practice involved tilling the land, meaning plowing. The new method is a no till method, meaning that organic remains of previous crop are left in the top soil. What this does is prevents the top soil erosion, which can occur when it is plowed and exposed to the wind. The no-till method delivers benefits in many situations including improved quality of crops and better retention of water in the soil due to the roots and stalks left from previous crop.

References:
http://www.MotherEarthNews.com
http://www.science.naturalnews.com
http://www.farming.co.uk
http://www.Webster'sOnlineDictionary.com.

Glyphosate

Glyphosate is more commonly known by its original name Roundup (manufactured by Monsanto), which is the world's most widely used herbicide. Glyphosate-based herbicides are manufactured by many different companies in many countries now that the patent period has expired. Glyphosate is the active ingredient of some of the most common herbicides used in farming and gardening. It is one of the most applied herbicides in areas such as roadsides, on footpaths, schoolyards, and sport fields. It is sprayed in the national parks and other environmental sensitive areas with the belief that it is nontoxic. It is widely sprayed on numerous crops and plantations. It is herbicide of choice, especially on so-called Roundup ready crops that have been genetically modified, such as corn, canola, cotton, soybeans, and sugar beets, to name a few. The result of glyphosate resistance is that glyphosate can then be applied without discrimination to area or dose. In the past, the use of glyphosate to control weeds had to be balanced with the cost of losing crops due to inadvertently heavy crop loss. These crops have been genetically modified to be herbicide resistant. Relatively high levels are permitted as residues in food and animal feed. Glyphosate spraying has dramatically increased with the introduction of these genetically modified crops. This also increases the risk of excessive occupational exposure as well as environmental contamination. Farmers are using it extensively, as well as general population for weed control. They do not take any precaution during application, believing the industry statement that it is safe. The facts and recent studies indicate that **this is not the case.**

Pure glyphosate has a low acute toxicity level. When it is combined with other ingredients, it is more toxic to plants. Studies indicate that the commercial product Roundup can be three times more toxic. Many deaths have occurred, especially in Asia. Widespread poisonings have occurred in Latin America as a result of aerial spraying on GM soybean and coca crops. Residues of sprays have been recorded as far away as ten kilometers from the sprayed site. The spraying has resulted in numerous animal deaths and also food crop losses.

Symptoms of poisoning commonly reported from unintentional exposure are **vomiting, abdominal pain, diarrhea, gastrointestinal infection. itchy or burning skin, skin rashes, infections, raised blisters, burning or weeping eyes, blurred vision, headache, fever, rapid heartbeat, palpitation, raised blood pressure, dizziness, chest pains, numbness, insomnia, depression, difficulty in breathing, respiratory infections, dry cough, impaired vision, drop in blood pressure, twitches and tics, muscle paralysis, peripheral neuropathy, loss of gross and fine motor skills, sweating, and severe fatigue.**

Japanese researchers analyzing suicides have found that drinking 200 ml of glyphosate was lethal. Those that survived had a range of health problems, namely intestinal pains, vomiting, erosion of gastrointestinal tract, excess fluid in the lungs, lung dysfunction, clouding of consciousness, kidney damage, low blood pressure, damage to the larynx, and destruction of red blood cells. Studies in Sweden have linked glyphosate exposure to hairy cell leukemia and non-Hodgkin's lymphoma. These cancers were extremely rare but are rapidly increasing in the Western world, up by 73% in the US A higher incidence of Parkinson's disease has been found among farmers using glyphosate herbicides. Recent studies have shown that exposure to glyphosate is associated to a range of reproductive effects in humans and other animal species. Studies in Ontario, Canada, indicated that a father's exposure to the herbicide glyphosate was linked to an increase in miscarriages and premature births among farm families. There was a report not too long ago by the Fox News medical reporter that infertility has increased in the US.

Some researchers have concluded that glyphosate and its commercial formulations clearly present a risk of carcinogenic, mutagenic, and reproductive effects on human cells. Numerous laboratory studies have shown that glyphosate in the Roundup formulation can be genotoxic and endocrine disrupting. One study summarizes these effects occurring at doses substantially lower than those used in agriculture or as permitted as residue.

The journal *Archives of Toxicology* published a shocking study showing that "Roundup is toxic to human DNA even when diluted to concentrations ***450-fold lower than used in agricultural applications.*** This effect could not have been anticipated from the known toxicological effects of glyphosate alone. The likely explanation is that the surfactants polyoxyethyleneamine within Roundup dramatically enhances the absorption of glyphosate into exposed human cells and tissue. If this is true, it speaks to a fundamental problem associated with toxicological risk assessment of agrichemicals in general. Namely, these assessments do not take into account the reality of **synergistic toxicologies, that is, the amplification of harm associated with multiple chemical exposures occurring simultaneously."** http://www.Greenmedinfo.com

The article further states, "Another study published that year in the journal Toxicology, revealed that the inert ingredients such as solvents, preservatives, surfactants and other added substances are anything but inactive. They in fact contribute to toxicity in a synergistic manner, and ethooxylated adjuvants in glyphosate-based herbicides were found to be active principles of human cell toxicity." The most disturbing of all, the researchers claim that cell damage and even cell death can occur at the residual levels found on Roundup treated crops, lawns, and gardens, where the herbicide is applied.

Studies have demonstrated that glyphosate causes genetic damage in human lymphocytes and liver cells. Other studies have shown that it causes oxidative stress, cell cycle dysfunction, and disruption of

RNA transcription. All this contributes to carcinogenicity. Exposure to glyphosate-based herbicides at even low concentrations may/can result in reproductive and hormonal problems, miscarriages, low birth weights, birth defects, and various cancers, especially hematological cancers such as non-Hodgkin's lymphoma, and numerous cancers, such as in the breast and prostate.

Birth defects are skyrocketing in agricultural centers in Argentina. Miscarriages, fertility problems, and abnormal fetal development are all problems that are on a massive increase in Argentina, where many are exposed to extensive spraying of herbicides. Eighteen million hectares in Argentina grow the genetically modified crop soybean on which three hundred million liters of pesticides are sprayed. In one village surrounded by soybean plantations, the rate of miscarriages is one hundred times the national average, where glyphosate herbicide is used. Dr. Mercado Vasquez, a neonatal specialist at the Children's Hospital in Cordoba, featured in the documentary film *People and Power – Argentina: Bad Seeds:* ***"I see new-born infants, many of whom are malformed. I have to tell parents that their children are dying because of these agricultural methods. In some areas in Argentina the primary cause of death for children less than one year old is malformation."***

The EPA standard for glyphosate in American water supplies is 0.7 ppm. In Europe, the maximum allowable level in water is 0.2 ppm. Organ damage in animals has occurred at levels as low as 0.1 ppm, and in the study on cells in liver, embryonic and placental, cell lines were adversely affected at above levels. It was earlier reported that GM corn contains 13 ppm of glyphosate, compared to zero for non-GM corn. Such level has a staggering implication where Americans eat on average of approximately 193 pounds of genetically modified food.

A German study looked at glyphosate's role in the rise of toxic botulism in cattle. This used to be extremely rare, but the incidence has become increasingly common over the past fifteen years. Researchers are now finding that the beneficial gut bacteria in both animals and humans are very sensitive to residual glyphosate levels. Dr. Huber and Dr. Stephanie Seneff explained that certain intestinal bacteria produce bacteriocins that are specifically directed against *C. botulism*, as well as other pathogens. According to the authors, lactic-acid-producing bacteria that help defend against *Clostridium* pathogens are destroyed by glyphosate. This suggests that the *C. botulism* associated diseases may be due to glyphosate tainted animal feed.

The beneficial bacteria in one's body, specifically the gut, are the body's first line of defense against diseases. As a matter of fact, 70 to 80% of our immune system is located in the gut. Glyphosate causes extreme disruption of the beneficial microbial function in the gut. What is worse, it disrupts the beneficial bacteria but allows the pathogens to overgrow and take over, resulting in inflammation leading to some of the diseases. Is there any wonder why so many diseases are on the increase?

Glyphosate is assumed by regulators to have no neurological effects. (One problem with word *assume* is when divided into three separate words, ass-u-me, meaning it makes an ass out of you and me.) The problem is that EPA did not require neurotoxicity studies to be performed for the registration of Roundup. There is emerging evidence that glyphosate can affect the nervous system and in particular areas of the brain associated with Parkinson's disease.

http://www.science.naturalnews.com
http://www.Globalsearch.org,Junr13,2013
http://www.Ecotoxicology.2013March.22(2);251-262
http://www.Greenmedinfo.comJune13,2013
http://www.Greenmedinfo.comOctober15,2012
http://www.ArchivesofToxicology2012May.86(5)805-13

Some of the misleading research methods used by biotech industry consist of the following steps:

1) Use of animals with varied weights to hinder the detection of food related changes.
2) Keep feeding studies short to miss long-term impacts.
3) Test Roundup ready soybeans that have never been sprayed with Roundup, as they always are in real world.

4) Avoid feeding animals the GM crop but instead give them a single dose of GM protein produced from GM bacteria.
5) Use too few subjects to obtain statistical significant level. **(Figures do not lie, but liars do figure.)**
6) Use poor or inappropriate statistical methods, or fail to even mention statistical methods or include essential data.
7) Employ insensitive detection techniques—doomed to fail.
8) Researchers tested GM soy on mature animals, not the more sensitive young ones. GM expert Arpad Pusztai stated, "Older animals would have to be emaciated or poisoned to show anything,"
9) Organs were never weighed.
10) The GM soy was diluted up to twelve times, which according to an expert review, "would probably ensure that any possible undesirable GM effect did not occur."
11) The amount of protein in the food was artificially high which would mask negative impacts of the soy.
12) Samples were pooled from different locations and conditions making it easily impossible for compositional differences to be scientifically significant.

Data from the only side-by-side comparison was removed from the study and never published. When it was recovered, it revealed that Monsanto GM soy had significantly lower levels of important constituents. The toasted GM soy meal had nearly twice the amount of a lectin—which interferes with the body's ability to assimilate nutrients. The amount of trypsin inhibitor, a known soy allergen, was as much as seven times higher in cooked GM soy compared to a cooked non-GM control soy. Monsanto named their study "The composition of glyphosate tolerant soybean seeds is equivalent to that of conventional soybean."

It is not a surprise that the four peer-reviewed animal feeding studies that were performed more or less in collaboration with private companies "reported no negative effects of the GM diet." On the other hand, adverse effects were reported but not explained in five different independent studies. It is remarkable that these effects have all been observed after feeding for only ten to fourteen days.

Monsanto's study on Roundup indicated that twenty-eight days after application only 2% of their herbicide had broken down. The weed killer is advertised as biodegradable, "leaves soil clean," and "respects the environment." These statements were declared false and illegal by judges in both the US and France. The company was forced to remove "biodegradable" from the label and pay a fine.

EPA claims their approval is based on a large body of science. It is a misguided approval process. The vast majority of research has major conflicts of interest, as it is done by company selling these GM crops and other pest- and weed-control chemicals. This was brought to the public's attention in a January 2014 report on 2,4-D, jointly published by Testbitech, GeneWatch, UK and Pesticide Action Network, Europe. Many of the publications are authored by the manufacturer's scientists or are sponsored by the manufacturers of component 2,4-D. For the reader's information, 2,4-D was an active component in Agent Orange, widely used to defiolate areas during Vietnam war.

http://articles.mercola.com/ . . . /articles/archive/2013/06/09/mosanto-roundup-herbicide.aspx.

http://www.momsacrossamerica.com/g;y[hosate_testing_results-190k

http://www.patft.uspto.gov/netacgi/nph-Parser http://www.archtatent.com/patents/3150632

Environmental Effects of Glyphosate

Glyphosate's existence has led scientists to develop Roundup ready seeds which are genetically modified to resist the glyphosate. This has spurred increase in engineered food. Farmers can plant these seeds and simultaneously spray the fields with glyphosate. Now farmers depend on the modified seed, creating a dependence on the chemical companies. This genetic modified food is being pushed on the third world countries under the guise of feeding the world.

The big problem is that the human body was intended to eat unmodified food. Independent studies have shown that glyphosate sinks into the human organs and tissues, where it accumulates creating a toxic environment for the human body.

http://www.rag,org.an/modifiedfoods/rounduphealthissues.htm,

http://www.beyondpesticides.orgdailynewsblog/?p=10487

Glyphosate herbicides have significant lethal effect on beneficial insect species on farms. About 50 to 80% of beneficial insects are killed by exposure to glyphosate herbicide. It is also very toxic to fish and other aquatic organisms. It is also highly toxic to organisms in the soil. It significantly reduces nitrogen-fixing bacteria, which transforms soil nitrogen to forms that plant use for growth. One application causes a dramatic change in decreasing beneficial soil micro-organisms and arthropods. This results in reduction of species that build humus, contributing to decline in organic matter. Numerous other studies indicate that glyphosate herbicide increase the susceptibility of plants to diseases. This is partly due to the reduction of growth of mycorrhizal fungi and other beneficial fungi that help the plant absorb nutrients and help fight disease. The other concern is the increase in soil pathogens, and decrease beneficial organisms that control disease. Glyphosate damages and reduces the population of earthworms. It also reduces populations of small mammals and birds due to damaging the vegetation that these animals use for food or shelter.

Glyphosate is a broad-spectrum herbicide, nonselective type herbicide that is transported throughout the plant causing damage. If it does not kill the plant, it can last for many years. Spray drift rates have been shown to damage surrounding vegetation. This should be a main reason to ban its use in national parks since it also kills wild plants.

References: Article by Carol Cox in *Journal of Pesticide Reform*, Fall 1998, vol. 18, no. 3; Hardell, L., and Eriksson, M. (1999), "A case study of Non-Hodgkin's Lymphoma and exposure to Pesticides," *Cancer*, vol. 85, no. 6 (March 15, 1999).

Health and Environmental Effects" A. Leu, May 2007

New Seralini study shows Roundup damages sperm

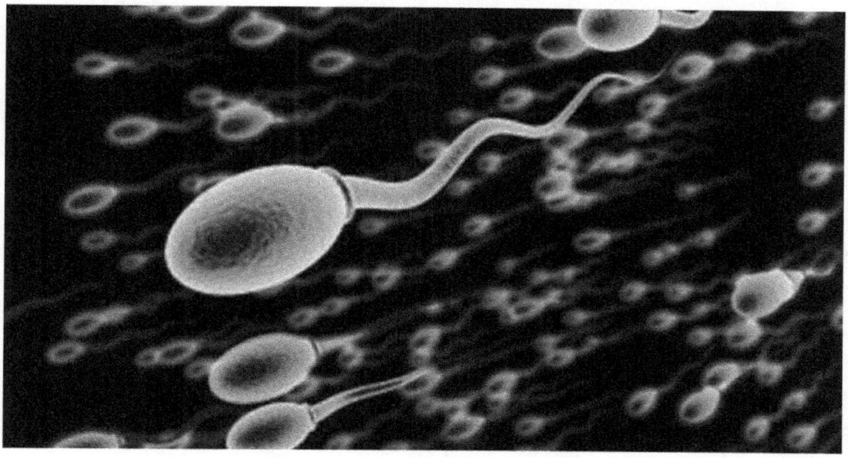

"A new study in rats found that Roundup altered testicular function after only eight days of exposure at a concentration of only 0.5%, similar to levels found in water after agricultural spraying," writes Claire Robinson, managing editor of *GMO Seralini*. The study found no difference in sperm concentration, viability, and mobility, but there was an increase in abnormal sperm formation measured two, three, and four months after this short exposure.

The study, the first to measure the delayed effects of exposure to Roundup on sperm in mammals from a short exposure, was conducted by a team including Prof. Gilles-Eric Seralini at the University of Caen, France. Roundup was found to change gene expression in sperm cells, which could alter the balance of the sex hormones androgen and estrogen. A negative impact on sperm quality was confirmed, raising questions about impaired sperm efficiency. The authors suggested that repeated exposures to Roundup at doses lower than those used in agriculture could damage mammalian reproduction over the long term.

The study's findings should raise alarm in farm workers, as well as people who spray Roundup for municipal authorities and even home gardeners. People exposed to lower doses repeated over the long term, including consumers who eat food produced with Roundup and people who happen to be exposed to others' spraying activities, should also be concerned. Those who want to conceive a child should take special measures to minimize their exposure, including eating organic food and lobbying for a ban on Roundup spraying in their neighborhoods.

An acute exposure to glyphosate-based herbicide alters aromatase levels in testis and sperm nuclear quality (Estelle Cassault-Meyer, Steeve Gress, Gilles-Éric Seralini, Isabelle Galeraud-Denis, *Environmental Toxicology and Pharmacology*, volume 38, issue 1, July 2014, pp. 131–140).

Highlights

- We investigated the effects of a glyphosate-based herbicide after an eight-day exposure of adult rats.
- We have shown a significant and differential expression of aromatase in testis.
- We have observed a diminution of mRNA expression of nuclear markers in spermatozoa.
- These results suggest changes in androgen/estrogen balance and in sperm nuclear quality.
- The repetition of exposures of this herbicide could alter the mammalian reproduction.

Copy of abstract: "Roundup is the major pesticide used in agriculture worldwide; it is a glyphosate-based herbicide. Its molecular effects are studied following an acute exposure (0.5%) of fifteen 60-day-old male rats during an 8-day period. Endocrine (aromatase, estrogen and androgen receptors, Gper1 in testicular and sperm mRNAs) and testicular functions (organ weights, sperm parameters and expression of the blood–testis barrier markers) were monitored at days 68, 87, and 122 after treatment, spermiogenesis and spermatogenesis. The major disruption is an increase of aromatase mRNA levels at least by 50% in treated rats at all times, as well as the aromatase protein. We have also shown a similar increase of Gper1 expression at day 122 and a light modification of BTB markers. A rise of abnormal sperm morphology and a decrease of the expression of protamine 1 and histone 1 testicular in epididymal sperm are observed despite a normal sperm concentration and motility" (reprinted from: http://www.gmoseralini.org/wp-content/uploads/2014/07sperm.jpg).

Additional Studies of Glyphosate Effects of Glyphosphate

Dr Santadino: Glyphosate Destroys Earthworm Eisenia fetida Populations

Using glyphosate herbicide can wipe out local earthworm populations, a new study shows.

By Henry Rowlands|November 27, 2014|Animal EvidenceRoundup EvidenceSouth America

Dr Herbert: Field-Realistic Doses of Glyphosate Lead to Honeybee Starvation

A new study published in the Journal of Experimental Biology establishes a link between the world's most sold herbicide Roundup and the dramatic decline in honeybee (Apis mellifera) populations in North American and Europe.

By Henry Rowlands|August 14, 2014|Animal Evidence, Roundup Evidence, South America

Dr Kru¨ger: Glyphosate Increases Birth Defects in Piglets

Glyphosate has been found in malformed piglets. The research study was conducted by a team of researchers from Germany and Egypt in collaboration with the Danish pig farmer Ib Pedersen, whose pigs were analysed for glyphosate content.

By Henry Rowlands|July 28, 2014|Animal Evidence, Europe, Lab Evidence, Roundup Evidence

Dr Gassmann: Western Corn Rootworm Resistant to Bt Toxin in GM Maize

According to a team led by Aaron Gassmann, an entomologist at Iowa State University in Ames, in some Iowa fields a type of beetle called the western corn rootworm (Diabrotica virgifera virgifera LeConte) has developed resistance to two of the read more

By Henry Rowlands|March 19, 2014|Animal Evidence, North America|0 Comments

Read More

Dr Krüger: Glyphosate is Toxic to Dairy Cows

A new study by a team of scientists led by Dr Monika Krüger has revealed that Glyphosate (Roundup) is toxic to the normal metabolism of dairy cows.

By Henry Rowlands|September 12, 2013|Animal Evidence, Europe, Roundup Evidence|3 Comments

Read More

Dr Huber: Glyphosate Could Cause Bee Colony Collapse Disorder (CCD)

There have been discussions about neonicotinoids, poor nutrition, Nosema, and mysterious viruses causing bee colony collapse disorder (CCD). Now plant pathologist Dr. Don Huber has pointed a finger at glyphosate as a possible cause of CCD in a paper

Read More

By Henry Rowlands|August 5, 2013|Animal Evidence, North America, Roundup Evidence|1

Read More

Dr Traavik: Low Levels of Roundup Toxic to Aquatic Invertebrate Daphnia Magna

Low levels of glyphosate based herbicide induced significant negative effects on the aquatic invertebrate Daphnia magna. Glyphosate herbicides such as brands of Roundup, are known to be toxic to daphnids.

Read More

By Henry Rowlands|July 3, 2013|Animal Evidence, Europe, Roundup Evidence|0 Comments

Read More

Dr. Tabashnik: Five Pest Species Immune to Bt poisons in GMO Corn and Cotton

As of 2010, five of 13 major pest species had become largely immune to the Bt poisons in GMO corn and cotton, compared to just one species in 2005, scientists report in this study.

By Henry Rowlands|June 14, 2013|Animal Evidence, North America|2 Comments

Read More

Dr. Judy Carman: Evidence of GMO Harm in Pig Study

A groundbreaking new study shows that pigs were harmed by the consumption of feed containing genetically modified (GM) crops.

By Henry Rowlands|June 12, 2013|Animal Evidence, Australia, Lab Evidence|0 Comments

Read More

Dr Krista B. Oke: GM Salmon Can Breed with Wild Trout

Genetically modified salmon can breed with wild trout to produce a new fast growing fish that can harm natural species, scientists have warned.

By Henry Rowlands|May 30, 2013|Animal Evidence, Europe, Lab Evidence|0 Comments

Latest Evidence

- Dr Santadino: Glyphosate Destroys Earthworm Eisenia fetida Populations
- Dr. Swanson: GM Crops and Glyphosate Linked to US Health Deterioration

- Glyphosate Frequently Detected in Soils and Water Across America
- Dr Herbert: Field-Realistic Doses of Glyphosate Lead to Honeybee Starvation
- Dr Kru¨ger: Glyphosate Increases Birth Defects in Piglets

Tags

bees birth defects Bt Corn Bt cotton Bt Maize cell death CRIIGEN death DNA Gilles-Eric Seralini Gilles-Eric Seralini research glyphosate glyphosate poisoning GM corn gm crops GM Free Cymru GM Maize GM maize pollen GMO GMOs GM soy gm soybean GM Wheat herbicide herbicides increase India infertility iron kidneys larvae liver mice MON810 Monsanto organ damage pesticides rats rhizosphere roundup Seralini research Sustainable Pulse toxic toxicity University of Caen

Latest GMO News

- German Poultry Industry Giant Returns to GMO-Free Production
- Roundup is toxic to the heart – new study
- Norwegian Authorities Ban GM Fish Feed over Antibiotic Resistance Fears
- Dr Santadino: Glyphosate Destroys Earthworm Eisenia fetida Populations
- Dr. Swanson: GM Crops and Glyphosate Linked to US Health Deterioration
- Culinary pleasures or hidden poisons?
- Coalition Steps in to Defend Maui Residents from Pesticides and GE Contamination

WHO Admits Monsanto's Glyphosate May Cause Cancer

On the latest news reported by Natural News is that the British Journal ***The Lancet Oncology*** published a groundbreaking report regarding glyphosate as "probably carcinogenic to humans," listing it as the second-highest classification for substances that could cause cancer. This classification is just below known carcinogens. The paper also listed abundant evidence of negative effect on human health by glyphosate. Glyphosate has been extensively found in agricultural workers' urine. Assessment of case control studies on occupational exposure to glyphosate conducted in Sweden, Canada, and the US, the researchers noted increased risks of non-Hodgkin's lymphoma. Other cancer types shown to develop due to glyphosate exposure include hemangiosarcoma, pancreatic islet-cell adenoma plus skin tumors. It was also noted that AMPA (aminomethylphosphoric acid), glyphosate breakdown product, was found in human blood.

The full report can be accessed at: http://wwwthelancet.com

So the question now comes up: What is EPA going to do now? The findings published in the *Lancet Oncology* journal contradict the official position of the US government. The EPA in 2013 upped the tolerance level for glyphosate in food crops. The EPA's position continues to state that the available evidence of glyphosate potential to induce cancer is inadequate. But at the same time, EPA does admit that glyphosate can cause major kidney damage in humans, inhibit normal reproduction, and cause lung congestion. Of course, Monsanto opposes these findings citing these findings are a departure from conclusions reached by regulatory agencies around the world. Researcher from The Institute for Responsible Technology, Jeffrey M. Smith, states, "Monsanto's takeover of FDA has been replicated around the world. I've been to 37 countries and I've seen how they 'capture' regulators, ministries, departments, etc., and once that happens, they discredit and dismiss any adverse findings about GMO's—they don't even read the dossier." This is a very real concern to all the populations of the world. Who wants to be slowly poisoned?

A study by the US Geological Survey, published in the journal *Environmental Toxicology and Chemistry* in 2014, found that in some areas, 75% of air and rain samples are contaminated with glyphosate.

http://thinkprogress.org
http://www.reuters.com
http://www.cnn.com
http://www.thedailysheeple.com
http://ecowatch.com

Glyphosate Toxic Mechanism on Harming Brain

A new study reveals a hitherto unknown mechanism behind how the world's most popular GMO herbicide harms the brain.

Remarkably, despite Roundup herbicide's widespread approval around the world, the most basic mechanisms through which it exerts toxicity toward non-target animal species (including humans) have yet to be adequately characterized. Concerned about Brazil's status as the largest global consumer of pesticides since 2008, researchers sought to elucidate toxicological effects of these agrochemicals in humans.

Their new study, published in the journal *Toxicology*, provides a proposed mechanism for the adverse neurological effects of Roundup (a glyphosate-based herbicide). It has been observed that agrochemical exposure can lead to, or accelerate, neurodegenerative disorders, such as Parkinson's and Alzheimer's disease. However, lacking a mechanism of action, such a link can more easily be written off as coincidental, which is largely the position of the medical establishment, agricultural industry, and its would-be regulators. The authors point out that "neurodegenerative conditions are frequently associated with glutamatergic excitotoxicity and oxidative stress," which is why they decided to investigate the subject further.

Titled "Mechanisms underlying the neurotoxicity induced by glyphosate-based herbicide in immature rat hippocampus: Involvement of glutamate excitotoxicity," the paper tested the neurotoxicity of Roundup in the hippocampus of immature rats following acute exposure (thirty minutes) and chronic (pregnancy and lactation) exposure.

The results found that acute exposure to Roundup induces calcium influx into neurons (primarily, by activating NMDA receptors and voltage-dependent Ca2+ channels), leading to oxidative stress and neural cell death. They also found that the herbicide affected the enzymes ERK and CaMKII, the latter of which is an enzyme whose dysregulation has been linked to Alzheimer's disease. Additionally, acute exposure was observed to have the following three effects: Increase the amino acid glutamate into the junctions through which neurons communicate (synaptic cleft), which, when released in excess levels, can exert excitotoxic/neurotoxic effects in neurons. (NMDA receptor is a brain receptor that is activated by the amino acid glutamate, which when excessively stimulated may cause cognitive defects in Alzheimer's disease. The name is N-methyl-d-aspartate receptor.)

Decrease the neuroprotective antioxidant glutathione

Increase "brain rancidity," i.e., lipoperoxidation, characterized by excitotoxicity (overstimulation of the neurons) and oxidative damage. The summarization of their results, looking at the effects of both acute and chronic exposure, was reported as follows: "Taken together, these results demonstrated that Roundup might lead to excessive extracellular glutamate levels and consequently to glutamate excitotoxicity and oxidative stress in rat hippocampus."

Roundup-induced glutamate excitotoxicity appears to share similar effects to monosodium glutamate (MSG) and aspartame-linked excitotoxicity, indicating that anyone either prone to, or suffering from, a brain or neurological condition involving increased oxidative stress and/or neuronal excitotoxicity (pathological or excessive nerve cell stimulation) should be even more wary to reduce exposure to this unfortunately ubiquitous environmental and food contaminant. The authors also pointed out that their study found maternal exposure to Roundup resulted in the offspring being exposed to the herbicide because it crosses the placental barrier during gestation and/or it is passed to them through the breast milk. They caution, "Exposure to environmental toxicants during pregnancy and suckling periods has the potential to affect embryo and fetal development." This is not the first time that concerns have been raised about Roundup's unique contraceptive and birth defect causing properties.

For additional information on the adverse effects of Roundup herbicide and related glyphosate formulations, visit the database sections on the topic, which references peer-reviewed and published research from the National Library of Medicine on the chemical Roundup herbicide, glyphosate formulations:

[i] Daiane Cattani, Vera Lúcia de Liz Oliveira Cavalli, Carla Elise Heinz Rieg, Juliana Tonietto Domingues, Tharine Dal-Cim, Carla Inês Tasca, Fátima Regina Mena Barreto Silva, Ariane Zamoner. MECHANISMS UNDERLYING THE NEUROTOXICITY INDUCED BY GLYPHOSATE-BASED HERBICIDE IN IMMATURE RAT HIPPOCAMPUS: INVOLVEMENT OF GLUTAMATE EXCITOTOXICITY. Toxicology. 2014 Mar 14. Epub 2014 Mar 14. PMID: 24636977

[ii] Yamauchi, Takashi (August 2005). "Neuronal Ca2+/calmodulin-dependent protein kinase II—discovery, progress in a quarter of a century, and perspective: implication for learning and memory." Biological and Pharmaceutical Bulletin 28 (8): 1342–54. doi:10.1248/bpb.28.1342. PMID 16079472.

Two Widely Used Fluoride Pesticides

Due to its high toxicity, fluoride has long been used as a pesticide. In the US, there are currently two fluoride pesticides that are allowed to be sprayed on food and in food processing plants. These are cryolite and sulfuryl fluoride.

Cryolite

Cryolite is a fluoride containing mineral that EPA permits to be used as a pesticide on thirty-two different food products. Cryolite leaves a toxic fluoride residue in and on fruits and vegetables. Chemical formula for cryolite is Na_3AlF_6, sodium aluminum fluoride. Initially it was mined but now we do have synthetic process. Cryolite's aluminum fluoride structure loses fluoride ions in the presence of water. The loose fluoride ions are very effective as pesticide active, but also leave a toxic residue on the food supply. It can be applied as powder or in liquid form. The most common uses of cryolite is on California grapes, potatoes and citrus.

EPA provides a list on the fruits and vegetables that cryolite can be used. These are apricots, beet roots, blackberries, blueberries (huckleberries), boysenberries, broccoli, brussels sprouts, cabbage, cauliflower, citrus fruits, collards, cranberries, cucumbers, dewberries, eggplant, grapes, kale, kohlrabi, lettuce, loganberries, melons, nectarines, peaches, peppers, plums (fresh prunes), pumpkins, radish roots, raspberries, rutabaga roots, squash (winter), squash (summer), strawberries, tomatoes, turnip roots, and youngberries.

California wines are high in fluoride since cryolite is the pesticide used. The residue on California wines range from 3 to 9 ppm/liter. The European Union has placed a restriction on importation of California wines at 3 ppm/liter, since has lowered to 1 ppm. Consider a 6 oz. glass of California wine (175 ml) containing 3-9 ppm/liter fluoride. It delivers a concentration that is 24 to 73 times higher than allowed in the municipal water system. Currently, the EPA is allowing 7 ppm on fruits and vegetables. There was a proposal to increase these levels on the following items: cabbage from 7 ppm to 45 ppm; citrus fruit from 7 ppm to 95 ppm; collards from 7 ppm to 35 ppm; eggplant from 7 ppm to 30 ppm; **lettuce from 7 ppm to 180 ppm**; leaf lettuce from 7 ppm to 40 ppm; peaches from 7 ppm to 10 ppm; raisins from 0 ppm to 55 ppm; tomatoes from 7 ppm to 30 ppm; tomato paste from 0 ppm to 45 ppm. These were proposed in 2011. **I guess the 7 ppm level does not poison us as fast enough, therefore the levels had to be increased to speed up the process.**

One can contact the EPA at the following address:

By mail: Product Manager (PM) 14, Registration Division (7505C), Office of Pesticide Programs, Environmental Protection Agency, 401 M St., SW., Washington, DC 20460. Office location, telephone number, and email address: Rm. 219, CM #2, 1921 Jefferson Davis Highway, Arlington, VA, (703) 305-6600, email: forrest.robert@epamail.epa.gov.

The effect it has on bugs is by killing them as a stomach poison. The effect on humans according to documented studies consists of anemia, neurological problems, endocrine disruption, skin rashes, bone issues, and stomach and intestinal problems. It is highly probable that it contributes to obesity, autism, ADD, ADHD, plus innumerable endocrinology ailments. The rate of these ailments is increasing rapidly from year to year.

http://www.fluoridededective.com/fluoride-facts/creolite

http://www.epa.gov/pesticide/factsheet/security

There are five important facts regarding fluoride exposure due to consumption of grape products. This is because the use of cryolite is widespread among US vineyards.

According to the USDA (2005) the average fluoride levels in grape product are as follows:

White grape juice	2.13 ppm.
White wine	2.02 ppm.
Red wine	1.05 ppm.
Raisins	2.34 pp.

Many juice drinks that are labeled as grape juice use grape juice as filler ingredient. The use of cryolite thus contaminates many fruit juices with fluoride.

Cryolite is also allowed to be added to the following products (although it is unclear how many producers actually do so, and what the resulting levels are): apricot, broccoli, brussels sprouts, cabbage, cauliflower, citrus fruit, collards, eggplant, kale, kiwi fruit, kohlrabi, lettuce, melon, nectarines, peach, pepper, plum, pumpkin, squash, (summer and winter), tomato, blackberry, blueberry (huckleberry), boysenberry, cranberry, dewberry, loganberry, raspberry, strawberry, and youngberry.

The key to avoid exposure to fluoride from cryolite is to avoid buying nonorganic grape products, particularly beverages made out of white grapes.

Sulfuryl fluoride, or trade name Vikane Gas

Sulfuryl fluoride, a fumigant, breaks down rapidly in the human body when one is exposed to it to its anion fluoride. It is exceptionally toxic, and workers who use it, are at risk. A study was done on the workers using sulfuryl fluoride. The study found that sulfuryl fluoride workers suffered subclinical effects on the central nervous system as well as cognitive deficits.

In 2005, the US Environmental Protection Agency (EPA) granted a request from DOW AgroScience to use sulfuryl fluoride as a fumigant in food processing plants as a means of killing bugs, rodents, and reptiles. FAN (Fluoride Action News) collaborated with the environmental group Beyond Pesticides to reverse EPA's approval through a series of substantive submission to the EPA. Although EPA granted FAN's request in January 2011 to rescind its approval of DOW's request, this as it currently stands, sulfuryl fluoride is still being sprayed on food products made in the US. The EPA allows sulfuryl fluoride as both a fumigant of food processing facilities (while food is still on the premises) and as a direct fumigant of food. Both forms of fumigation result in contamination of food with fluoride. Here is what one needs to know about both:

Fumigation of food processing plants

1) Structural fumigation is done for the purpose of killing pests in the facility where the food is stored. It usually is performed twice a year.
2) Unlike virtually every other Western country, the EPA does not require that food processors remove food prior to the fumigation. As a result, any food that is still being stored in the facility during structural fumigation will be contaminated with fluoride. Western countries require that the food be removed from the facility prior to fumigation, clear the facility of the fumigate, and discard the first 10% of processed food after startup. I guess no one is interested in keeping us healthy.

3) The level of fluoride contamination that EPA allows for wheat flour (125 ppm) and dried eggs (900 ppm) is sufficient to cause symptoms of acute fluoride poisoning (e.g., nausea, vomiting, etc.) in children.
4) Although less than 0.1% of wheat flour and dried eggs will be contaminated with sulfuryl fluoride (due to the infrequency of structural fumigations), several hundred, if not thousands of children will be exposed in a year to doses of fluoride from these products that can induce temporary food poisoning type symptoms. No other country does this.
5) There are hundreds of other food products that EPA allows to be contaminated with sulfuryl fluoride.

Direct fumigation of food

1) The EPA also allows food processors to use sulfuryl fluoride as a direct fumigant of certain foods. This means that food processors can purposely spray sulfuryl fluoride directly onto certain foods.
2) Unlike structural fumigation, which takes place once or twice a year, direct fumigation is a routinely performed procedure. Foods that can be directly fumigated with sulfuryl fluoride will consistently have elevated fluoride levels.
3) According to EPA's estimates, some foods that will be the most commonly fumigated are cocoa powder, dried beans, walnuts, and dried fruits.
4) EPA estimates that if the current regulations are not rescinded, 100% of cocoa powder, 100% of dried beans, 99% of walnuts, 69% of dried fruits, 10% of tree nuts, and 3% of brown rice will be fumigated.
5) When fumigated, the average fluoride content in fumigated foods will be as follows:

Brown rice	12.5 ppm
Cocoa powder	8.4 ppm

Almonds	5.3 ppm
Tree nuts	5.3 ppm
Dried beans	4.5 ppm
White rice	4.5 ppm
Walnuts	2.4 ppm
Dried fruits	1.0 ppm

Here is an example what one consumes from dried eggs which are most likely used as a direct replacement for fresh eggs in recipes like scrambled eggs or omelets.

Fluoride residue in dried eggs is 900 ppm where 900 ppm = 900 mg/kg. Average weight of one fresh egg = 50 grams (American Egg Board 2005). Conversion factor from dried egg to fresh egg is 1 part of dried egg to 3 parts by weight of water (American Egg Board 2005)

12.5 grams dried egg mixed with 37.5 grams of water gives 50 grams of reconstituted egg.

12.5 grams × 900 mg/kg × 0.001 kg/g = 11.25 mg per egg of fluoride

2 egg equivalent 2 × 11.25 mg = 22.5 mg of fluoride

4 egg equivalent 4 × 11.25 mg = 45 mg of fluoride

The recommended water fluoridation level is 0.7–1.2 ppm/liter. How do the dried eggs stack up against the recommended levels?

http://www.fluoridealert/content/sf_exposure.html

The question arises: Is EPA really protecting our health or the financial rewards of the agricultural complex? Is there a more sinister plot to this?

This is an outrage. We should support organizations like the Fluoride Action News and petition the EPA to stop food fumigation.

References: Zhia, Z., Laine, P. L., Nicovich, J. M., Wine, P. H., 2010, "Reactive and non-reactive quenching of O(1D) by the potent greenhouse gases SO2F2, NF3, and SFSCF3," Proc Natl. Acad. Sci, USA. 2010 April 13; 107(15)6610-5. Full article; Andersen, M. P., Blake, D. R., Rowland, F. S., Hurley, M. D., Wellington, T. J., 2009, "Atmospheric Chemistry of sulfuryl fluoride; reaction with OH radicals, Cl atoms and O3 atmospheric lifetime, IR spectronal and global warming potential." Environmental Sci. Tech. 2009. Feb. 15; 43(5)1067=1070. FAN objections and EPA notices.

Neonicotinoid Pesticides

There is a new class of pesticides called neonicotinoids. These pesticides are water soluble, systemic, and pervasive effect on the insect's central nervous system, cumulative and irreversible.

The toxicity of neonicotinoid pesticide is similar to natural insecticide nicotine. They act on the nervous system, in such a manner as to excite the nerves to such a level that it causes paralysis, leading to death. All neonicotinoid pesticides are classified as general use and have been registered under EPA's Conventional Reduced Risk Program due to their favorable toxicological profiles. This committee set up within the EPA agency is for the sole purpose to expedite the review and regulatory decision-making process of conventional pesticides that pose less risk to human health and the environment than existing conventional alternatives. The goal is to quickly register commercially viable alternatives to riskier pesticides such as neurotoxins, carcinogens, reproductive and developmental toxicants, and ground water contaminants. The advantages to the chemical companies are an expedited decision timeframe. The companies are not allowed to put a reduced risk pesticide claim on their labels. The reduced risk pesticide status gives the companies a marketing advantage. http://www2.epa.gov:PesticideRegistration

The five neonicotinoid pesticides are (1) acetamiprid, (2) clothianidin, (3) imidacloprid, (4) thiamethoxam, and (5) dinotefuran.

Acetamiprid is used against sucking insects such as aphids and whiteflies. It is used on leafy vegetables, cole crops, citrus, cotton, ornaments, and fruiting vegetables. Clothianidin is used on corn and canola seed treatment. It is also used on grapes, pome fruit, tobacco, and turf and ornaments. Imidacloprid is the most widely used neonicotinoid pesticide. It is widely applied against a range of target pests and sites that include soil, seed, structures, pets, and foliar treatment on cotton, rice, cereal, peanuts, potatoes, vegetables, pome fruits, pecans, and turf. It has long residual activity and very effective against sucking insects, turf insects, and Colorado potato beetle. It is usually available as dust, granules, seed dressing as a flowable slurry concentrate, soluble concentrate, and wettable powders. Thiamethoxam neonicotinoid pesticide is the most water soluble, which means it readily moves in plant tissue. The product is used for soil and foliar treatment against a wide range of vegetable field crops. Product comes as emulsifier concentrate, water-disposable granules, and soluble concentrate. Dinotefuran pesticide acts through ingestion and within several hours due to cessation of feeding results in death shortly after.

Neonicotinoid pesticides are classified by EPA as toxicity class II and class III. The most available toxicity data is on imidacloprid. That indicates that it is less toxic. Animal studies on rats indicated such effects as lethargy, respiratory disturbance, occasional trembling, and spasm. No accounts of human poisoning, but toxic reaction could possibly be the same as in rats. The neonicotinoids, specifically imidacloprid, is the most toxic to birds and fish. The neonicotinoid pesticides are highly toxic to honeybees.

These pesticides weaken the bee's immune system. Forager bees bring the pesticide-laden pollen back into the hive, where it is consumed by other bees. Six months later, their immune system fails, falling prey to parasites, mites, viruses, and bacteria.

Independent review by numerous scientists from International Union for the Conservation of Nature examined eight hundred studies, and found that neonicotinoids insecticides are indeed gravely harming bees and other pollinators such as butterflies. The research also indicated harm to birds, earthworms, snails, and their invertebrates. Researcher Jean-Marc Bonmatin, member of the National Center for Scientific Research, stated, "The evidence is very clear. We are witnessing a threat to the productivity of our natural and farmed environment equivalent to that posed by organophosphates or DDT. Far from protecting food production, the use of neonicotinoid insecticides is threatening the very infrastructure which enables it."

The US Fish and Wildlife Service has announced they would be banning all neonicotinoid insecticides from wildlife refuges across the US by January 2016. The agency also announced it would phase out the use of genetically engineered (GE) crops to feed the wildlife in refugees, which is an environmental victory. **How about us humans?**

The second harm that occurs from neonicotinoid pesticide usage is that it is soluble in water. Sampling Midwestern stream sites in 2013, neonicotinoids were detected at all sites. This is reported by Mother Jones, "These findings directly contradict industry talking points. Older insecticides were typically sprayed onto crops in the field, while neonics are applied directly to seeds, and then taken up by the stalks, leaves, pollen, and nectar of the resulting plants. "Due to its precise application directly to the seed, which is planted below the soil surface, seed treatment reduces potential off target exposure to plants and animals—Croplife America, the pesticide industry's main lobbying outfit, declared in a 2014 report. Yet the USGS researchers report that pesticides that once rained down on the corn/soy belt, like chlorpyrifes and carbofuran, turned up at a substantially lower levels in water—typically in less than 20% of samples, compared to 100% of samples found in current neonic study."

"Apparently, pesticides that are taken up by plants through seed treatments, don't stay in the plants, and neonics," the USGS authors say, "are highly water soluble and break down in water slowly then the pesticides they've replaced" (http://www.edis.ifas.ufl.edu/pi07).

Other leading theories regarding bee die-offs are:

1) Pesticides, insecticides, and fungicides. Neonicotinoids such as imidacloprid, clothianidin, and sulfoxaflor kills insect by attacking their nervous system.
2) Malnutrition/nutrition deficiencies. Many beekeepers place the hives near fields of identical crops, which results in malnutrition, since bees get only one type of nectar. They need a wide variety of diversified nectar for their proper immune system function.
3) Viruses and fungi due to weakened immune system.
4) Electromagnetic fields (EMFs). Researchers have discovered that when a cell phone is placed near a hive, the radiation generated is enough to prevent bees from returning

to them, a study conducted by Landau University some years ago. A study in 2011 determined that microwaves from cell phones have a dramatic effect on bees, causing them to become distorted.

5) Lack of natural foraging areas due to mass conversion of grasslands to corn and soy fields. (monoculture).
6) Genetically modified (GM) crops. German study demonstrated that horizontal gene transfer appears to take place between the GM crop and the bees that fed on it. Bees released in a field of GM modified rapeseed, then fed the pollen to younger bees, the scientists discovered the bacteria in the guts of the young ones mirrored the same genetic traits found in the GM crops.
http://www.articles.mercola.com/ . . . /2013/08/18/neonicotinoid-pesticide.aspx,
http://www.wikipedia.org/wiki/Neonicotinoid

Biological Pesticide Bt

Bt toxin is a form of a biological pesticide—so what is it? A bacteria—*Bacillus thuringiensis*—is a naturally occurring bacteria in the soil. It has been used as an insecticide for a number of years. Once eaten by the insects, it attacks the gut of the insect and splits it open. It kills the insect that way, since they cannot eat. Monsanto decided to splice a gene from the Bt to corn and other plants like cotton, and soybeans. The company jointly claimed that the toxin would be destroyed in the digestive tract of mammals. The Bt toxin, therefore, is a biological pesticide. This Bt toxin was specifically added to corn against the pest corn borer. Following is a pictorial representation as provided by Dr. Eva Sirinathsinghji:

bacillus thuringiensis
Bt Gene is inserted into crop
Crop is infected by European corn borer
Pest dies when feeding on any plant part

A study published in the journal *Reproductive Toxicology* by doctors at Sherbrooke University Hospital, in Quebec, Canada, found that Bt toxin was found in the blood of pregnant women and their babies and in nonpregnant women as well. This study disproved the safety claims by the company and EPA. The executive director of the Institute for Responsible Technology, Mr. Jeffrey M. Smith, suspects that "the rise in Bt levels is due to another reason: colonies of the bacteria are replicating within the intestines, long after the GM food was ingested." If this hypothesis is correct then there will be an increase in gastrointestinal diseases, autoimmune diseases, food allergens, and childhood learning disorders. According to physicians, they are indeed seeing this increase.

Bt can be sprayed on crops or it is added to the DNA of a genetically modified food crop. It is widely used in the US, since it can be cheaply produced. Bt toxin is an aggressive toxin that damages cell membranes in the digestive tract. This phenomenon caused researchers to examine this negative effect on humans. The Bt aggressive mechanism harms the digestive system by creating so-called leaky gut. Research has indicated that Bt toxin persists in the gastrointestinal tract of human flora associated rats for weeks after exposure.

Bt exposure is of major concern for agricultural workers. A study conducted on workers handling Bt, it was found that almost 50% of them had Bt in their feces. It can cause serious lung damage and irritate airways upon inhalation during the dispersal. It is also found in the food we eat. It has been demonstrated that Bt insecticide residue can remain on fruits and vegetables. and it cannot be washed off when present in GM food. In 2006, it was found in pasteurized milk, ice cream, and juice, indicating that it stays through multiple stages of food production. Bt is the most widely used bio-pesticide in the agricultural industry; therefore, it is in the food we eat.

Bt toxin does interact with the intestinal walls of mammals, therefore with humans, as was demonstrated by a study on mice as well as the rhesus monkeys. If the Bt toxin is potentially causing holes in the intestinal walls of newborns, this results in toxins entering the blood stream. Some scientists speculate that this would lead to autoimmune diseases and food allergies. Could this be the cause of so many young people being allergic to peanuts and peanut butter to such an extent as to cause death? How about these toxins passing through the blood-brain barrier and causing cognitive problems, and even autism, which are all on an increase.

Other reports that came from Indian farmers using Bt toxin spray suffered from rashes and itching. Similar results were reported by about five hundred people in the state of Washington and Vancouver, complaining of flu-like symptoms and allergic reactions. Thousands of Indian farmers suffer similar reactions after using genetically modified cotton plants that produce Bt toxin. Another report that should be of a very important concern is that buffaloes in one Indian village that fed on Bt cotton plants, all thirteen died. Also, when their livestock was allowed to feed on Bt cotton plants after harvest, a large number of sheep, goats, and buffalo died. Yet when these animals grazed on natural cotton plants, they did not suffer any adverse health effects.

The Bt toxin produced in GM has known allergens that fail all three GM allergy tests recommended by the World Health

Organization: similarity of its amino acid sequence to known allergens, digestive stability, and heat stability. A specific Bt toxin found in Monsanto's yield guard and Syngenta's eleven Bt varieties is Cry1AB. It was discovered by an FDA researcher that the Cry1AB Bt toxin shared a sequence of nine to twelve amino acids with vitellogenin, an egg allergen. It is also very resistant to digestion and heat. This particular Bt toxin was produced by StarLink corn, grown for use as animal feed. It contaminated the US food supply in 2000, resulting in health complaints to the food manufacturers. After this incident, expert advisors to the EPA called for "surveillance and clinical assessment of exposed individuals to confirm the allergenicity of Bt products." No such program was started to date.

Another source of Bt toxicity comes from a cornfield that is pollinating. Such incident happened in a Filipino village where the entire population of that village came down with following symptoms: headache, dizziness, extreme stomach pain, vomiting, chest pain, fever and allergies, respiratory, intestinal, and skin reactions. The health effects repeated when the same GM corn was planted in another four villages and the same adverse health effects were observed at the time of cornfield pollination. Allergic reactions are a defensive, often harmful immune system response to an external irritant. There are over thirty scientific publications indicating harmful effects from the pesticide. Instead pursuing additional safety scientific studies, these scientists who publish data that happens to go against the safety of biotech products are under immense pressure from GM proponents and industry and even regulatory bodies. Why is that?

Now it has happened—some scientists have predicted the hazard of biotech products. The latest report now confirms what some scientists feared. The scientists have documented the rapid evolution of corn rootworm that are resistant to Bt corn. Seems like Mother Nature can counter with evolution to man's grandest agricultural solutions. Most of man's artificial solutions are of short timeframe before Mother Nature comes back with her own solution. The problem is that in man's agricultural solutions, he manages to destroy the ecosystem and potentially the future of mankind.

http://www.permaculturenews.org/ . . . /bt_toxicity_confirmed_flawed_studies_exposed

http://www.naturalnews.com/037405_bt-toxin_genetically_modified_food_detox.html#ixzz3a8H4g7Cl

Chapter 14

Vaccinations

This is a highly controversial subject with no lengthy studies ever published as to the safety of this medical procedure. It is almost approved just on the medical profession say so. There has been a multitude reports regarding vaccines. I decided to get more information on this subject matter.

To my pleasant surprise, there is quite a bit of information regarding vaccinations. The big question regarding vaccines is the belief that vaccinations are safe. The government, CDC (Centers for Disease Control), numerous prominent health officials, and of course the medical profession all recommend vaccination. The problem that I have uncovered as an answer to my question: Are there any safety studies done prior to releasing the vaccines for general use? And the answer is **no**. Then the questions begs, and why not? But that is true regarding recent vaccinations. Sixty to seventy years ago, vaccinations were quite beneficial. This no longer is true. Present-day vaccines contain quite a few so-called adjuvants, compounds to increase immune response to a given toxoid. Finally studies were done by some conscientious scientists as to the content in vaccines to determine their safety.

Dr. Robert Rowen has been investigating the claims regarding safety and effectiveness of vaccines for a good number of years. His undeniable conclusion is that vaccines *do not work* and they are *not safe*. According to Dr. Rowen's conclusion, there is a wide misconception that vaccines are the only way to attain immunity and avoid infectious disease. But quite the contrary is actually true, as vaccines provide only temporary immunity at best. Vaccines simultaneously and significantly increase the risk of immune dysfunction, behavioral disorders, and other major health problems. According to Dr. Rowen, "There is a graph of how these communicable diseases have fallen since the introduction of vaccines and a corresponding parallel, identical rise in chronic immune dysfunction like asthma, arthritis, multiple sclerosis, and others. No one has ever done an all-cause morbidity and mortality study on the effectiveness and safety of vaccines." A study published in the *Journal of the American Medical Association* (JAMA) back in 2010 highlights a doubling in the rate of chronic conditions among children between the years of 1994 and 2006—from 12.8% to 26.6%. This directly corresponds with a substantial increase in the number of vaccines added to the government's vaccine schedule, which included thimerosal, (ethylmercury) a known neurotoxin. Another statistic mentioned by Dr. Rowen is that mortality rate is higher among vaccinated children.

http://www.naturalnews.com/040007_Robert_Rowen_Vaccine_World_Summit_immunization.html#jxzz3Uh21Cs,

http://vaccineworldsummit.com,

http://vaccineliberationarmy.com

There are two types of vaccines. Some vaccines may contain a weakened or inactivated virus. If that virus stimulates a strong immune response, then usually no adjuvants are added. Vaccines made from weakened or killed pathogens contain naturally occurring adjuvants and can elicit potent protective immune responses. The second type is vaccines that contain adjuvants. Most vaccines developed today include small components of microbes, such as protein, rather than the entire virus or bacterium. These vaccines are necessarily formulated with adjuvants to generate a strong, long-lasting protective immune response. This is done to enhance the immune response to a given toxoid. The reason given is that pure recombinant or synthetic antigens used in modern vaccines are generally far less immunogenic than the older style live or killed whole organism vaccines. To improve the immune response, adjuvants were added to them. An adjuvant is a component that potentiates the immune responses to an antigen. It modulates the desired immune response. It is a substance that accelerates, prolongs, and enhances antigen-specific immune response. There are many known adjuvants that are in widespread use in vaccines. So what are some of the adjuvants used in vaccines?

The vaccine ingredients can be split into two categories:

Active Ingredients

These are aluminum, thimerosal (ethylmercury), gelatin, human serum albumin, sorbitol, emulsifiers, and taste improvements.

Products used in vaccine manufacture

Antibiotics, egg protein, yeast proteins, latex (in packaging), formaldehyde, glutaraldehyde, acidity regulators, human cell-lines, animal cell-lines, GMOs, and some other growing medium.

Thimerosal

This adjuvant was one of the more popular adjuvants added until 2006. Due to complaints of Moms Across America regarding autism, it was discontinued in the infant vaccinations. The content of thimerosal- a component containing mercury in the vaccine contributed to the autism and ADHD type neurological problems in children. Children were vaccinated at an early stage in life. Their brain and numerous organs are still developing. Injecting vaccines containing thimerosal is one of the ingredients that could have an adverse effect on their developing brain.

One of the more interesting informative article was revealed by CDC this past year that the 2014 flu vaccine was ineffective. The powers that be miscalculated as to the dominant flu strain for this particular year. Yet CDC was still encouraging people to get vaccinated. What for? I guess so that the pharmaceutical company that invested into this vaccine would not suffer a financial loss. But there also could be an ulterior motive—more mercury to be injected into the population to make them sick? The problem with this is that an analysis by an independent laboratory for heavy metals of this influenza vaccine documented mercury presence at **51 parts per million,** which is over twenty-five thousand times higher than the maximum contaminant level of mercury in drinking water set by EPA. The extremely high level of mercury in this flu shot was higher than found in tuna. This analysis for mercury found in this flu shot was according to this laboratory: **"100 times higher than the highest level of mercury we've ever tested in contaminated fish."** Other metals found in this vaccine were Aluminum at 0.4 ppm, beside the 51 ppm of mercury. What is really bothersome is that the doctors, pharmacists and members of the mainstream media continue to stage a lie, claiming mercury has been removed from vaccines.

http://www.naturalnews.com/045418_flu_shots_influenza_vaccines_mercury.html#ixzz3UjxHlt1s

The interesting listing on the insert for this flu vaccine was that no scientific trials were ever done. According to the insert, **"There have been no controlled trials adequately demonstrating a decrease in influenza disease after vaccination with Flulaval."** It also states, **"Safety and effectiveness of Flulaval have not been established in pregnant women, nursing mothers or children."** Or this statement: **"Safety and effectiveness of Flulaval in pediatric patients have not been established."** The insert further states that each dose of Flulaval contains up to 25 mcg of formaldehyde (a neurotoxin) and up to 50 mcg of sodium deoxycholate.

The mercury injected directly into the body is absorbed a 100%. Mercury ingested sticks to the food fibers and it is transported out of one's body. As you can gather from this information, influenza vaccines can cause a huge number of devastating health issues. So how can one believe the massive disinformation campaign practiced by the mainstream media, medical professionals, and government propaganda? Employees of some healthcare organizations being threatened with loss of employment if they do not succumb to vaccination, meaning to be poisoned.

Now there are multitude reports regarding the toxicity of mercury. Mercury has been linked as a contributor to autism, plus other neurological problems. Thimerosal was developed by Eli Lilly in 1930. The company tested thimerosal on twenty-two patients with terminal meningitis, who died weeks after being injected. This fact was never reported by the company in its study and declared thimerosal safe.

Researchers at Pittman-Moore in 1935 warned that thimerosal was not safe. They declared thimerosal as "unsatisfactory as a serum intended for use on dogs." It was widely used as a serum on soldiers being vaccinated during World War II, but it had to be labeled as a poison. A study published in *Applied Microbiology* found that thimerosal killed mice when added to injected vaccines. Even Lilly's studies indicated that thimerosal was "toxic to tissue cells" in concentrations as low as 1 ppm, which is a one hundred times weaker in a typical vaccine. Lilly went further and incorporated thimerosal into topical disinfectants. A report from a Toronto hospital showed that ten babies died when an antiseptic preserved with thimerosal was applied onto their umbilical cord. A school nurse, Patti White, told the House Government Reform Committee in 1999, "Vaccines are supposed to be making us healthier. However, in 25 years of nursing I have never seen so many damaged, sick kids. Something very, very wrong is happening to our children."

Institute of Medicine was requested to conduct a study ordering the researchers to "rule out" the chemical's link to autism. An epidemiologist by the name of Tom Vestraeten analyzed the agency's massive database containing the medical records of one hundred thousand children. A mercury-based preservative in vaccines—thimerosal—appeared to be responsible for a dramatic increase in autism and a host of other neurological disorders among children. He stated, "I was stunned by what I saw." The other findings from that study indicated a link between thimerosal and speech delays, attention-deficit disorder, hyperactivity, and autism. In 1991, CDC and FDA recommended three additional vaccines laced with thimerosal to be administered to very young infants. The estimated number of cases in autism had increased fifteen fold, from 1 in 2,500 children to 1 in 166. It is now predicted to be 1 in 44, increasing in occurrence.

Dr. George Lucier, a toxicologist and former director of the Environmental Toxicology Program at the National Institute of Health Sciences (NIEHS), clearly indicated that thimerosal, which is ethylmercury, "is a developmental neurotoxicant and exposure to it holds the same dangers as methylmercury." Dr. Lucier coordinated numerous toxicology research for a number of federal agencies like FDA, EPA, OSHA, and CDC.

These studies established that ethylmercury (thimerosal), the supposedly safe form of mercury according to vaccine establishments, converts in the rats and apparently in human infants into methylmercury. Ethylmercuric chloride, a material used previously as a fungicide, and banned long ago, is used to make thimerosal. This type of mercury deposits twice as much inorganic mercury in the brains of primates as compared to equal doses of methylmercury.

Health officials often claim that ethylmercury should not be compared to methylmercury, a known neurotoxin. A Russian study found that adults exposed to much lower concentrations of ethylmercury than those given to American children still suffered brain damage years later. National Institute of Health (NIH) study also found that ethylmercury is more toxic to the brain than methylmercury. It can cross the blood-brain barrier quicker and is converted to inorganic mercury. This form stays in the brain longer and at much higher concentrations. A Magos study in 1985 in the *Archives of Toxicology* had these key points listed: **"Neurotoxicity of methyl and ethylmercury were similar, although higher levels of inorganic mercury were seen in the brains of ethylmercury treated rats consistent with what we'd said about metabolism; and likewise, because of that, the renal damage was greater in the ethylmercury treated rats."**

http://www.ageofautism.com/2012/01/president-obama-i-will-not-back-down-from-protecting-our-kids-from-mercury-poisoning.html

http://www.whale.to/vaccine/ethy;_vs_methyl.html-9k,

http://www.decodedscience.com/ethyl-methyl-mercury-difference/25230-68k,

http://www.hindawi.com/journals/jt/2012/373678/-126k

Recommend reading a book titled *Evidence of Harm: Mercury in Vaccines and the Autism Epidemic*, by David Kirby. It contains 436 scientific references.

Dr. Russell Blaylock, a known neurosurgeon, warned, "The flu vaccine is completely worthless." Flu vaccines contain mercury in the form of thimerosal (ethylmercury), a brain toxin, which accumulates in the brain and other organs. It is incorporated into the brain for a lifetime. After five or ten years of flu shots, enough mercury accumulates in the brain that every single study agrees it is neurotoxic. Mercury is extremely toxic to the brain even in very small concentrations, and there are thousands of studies that prove it." The changes that are observed in the brain associated with neurodegenarative diseases such as Alzheimer's and Parkinson's are all easily produced by mercury in these doses.

Since 2006 or so, thimerosal has been removed from infant vaccines, although it may still be as trace amounts depending on the preparation of the vaccine. It is still extensively used in the flu vaccines, and as everyone is aware, the misleading news media, healthcare organizations, medical professionals, and CDC are still propagating that the public get their flu shots.

Aluminum

The second most abundant adjuvant extensively used in vaccines is aluminum. In many vaccines, certain aluminum salts serve as an immune adjuvant (immune response booster) to allow the protein in the vaccine to achieve sufficient potency as an immune stimulant. High concentrations of neurotoxic aluminum were added to child vaccines when mercury was removed. Two-month-old babies receive up to 1,225 micrograms (1,225 ppm) of aluminum in their vaccines, which is fifty times higher than safety levels. Aluminum toxicity has not received the widespread media attention that mercury has. Dr. John Clements, WHO vaccine advisor, stated, "Aluminum is not perceived, I believe, by the public as a dangerous metal. Therefore, we are in a much comfortable wicket in terms of defending its presence in vaccines." So how was the safety limit established of 4 to 5 mcg?

A study reported in *New England Journal of Medicine* (1997) compared the neurological development of about one hundred premature babies. One group were fed an intravenous nutrient containing 50 mcg per day of aluminum, and another group were fed with 10 mcg per day or 4 to 5 mcg of aluminum per kg of body weight.

The reason for this study was the knowledge that aluminum can build up to toxic levels in the bloodstream, bones, and brain. The infants who were given IV solutions with the higher levels of aluminum showed impaired neurologic and mental development at eighteen months, compared to infants with much lower amounts of aluminum. The interesting phenomenon is that FDA does not require aluminum warning on vaccines, but they are required in all other injectable medications. These warnings seem to be applicable only to premature babies and patients with kidney problems.

So one wonders how much aluminum is in the vaccines? According to Dr. Robert Sear's article on vaccine FAQs:

- Hib (PedVaxHib brand only—225 micrograms
- Hepatitis B—250 micrograms
- DTaP—170 to 625 micrograms, depending on the manufacturer
- Pneumococcus—125 micrograms
- HPV (Gardasil)—225 micrograms
- HPV (Gardasil-9)—500 micrograms
- Pentacel (DTaP, HiB, and polio combo vaccine)—330 micrograms
- Pediarix (DTaP, Hep B, and polio combo vaccine)—850 micrograms

Let's examine a typical injection schedule for infants. According to Dr. Sears, "A newborn gets a Hepatitis B injection on day one of life would get 250 micrograms of aluminum. This would be repeated at one month of age with the next Hep B shot. When a baby gets the first big round of shots at two months, the total dose of aluminum can vary from 295 micrograms (if a non-aluminum HiB and the lowest aluminum brand of DTaP is used) to a whopping 1225 micrograms if the highest aluminum brands are used and HiB B vaccine is also given. These doses are repeated at 4 and 6 months. A child would continue to get some aluminum throughout the first two years with most rounds of shots." The number of vaccines children received in 1950 was four in their first eighteen months of life. Now, they receive about seventeen. This translates on amount of aluminum received was about 1,120 mcg of aluminum. Now it is about 6,150 mcg of aluminum when the children get the recommended vaccination by the time they are two years old. The aluminum is 100% absorbed since it is given intramuscularly.

One should bear in mind that FDA cautions that premature babies and patients with impaired kidney function should not exceed more than 10 to 25 micrograms of injected aluminum at any one time. Then what about our infants? The numerous injections containing aluminum they receive by the time they are two years old. No wonder there so many children that have numerous health problems and potentially leading mental impairment such as ADHD, ASD, depression, etc. The aluminum content in vaccines reinforces the adverse effects of fluoride, aspartame, MSG, and GMO foods and drinks. The FDA and AAP (American Association of Pediatrics) documents state that it may be a problem, but they have not studied it. There are no reports or discussions regarding the levels of aluminum in vaccines, except by a few doctors that held a vaccine summit meeting in April 2015. The increase in aluminum content came about when thimerosal was removed from the infant vaccines starting in 2006. It is interesting to note that no AMA, Pediatrics or any numerous reps from CDC or FDA attended that summit meeting on vaccine safety.

There was one group that did a half-hearted analysis of aluminum content in vaccines. The researchers from Cochrane Collaboration (this group looks at healthcare problems around the world) investigated aluminum in vaccines and published their results in the *Lancet*

Infectious Diseases in 2004. This was well prior to the increase of aluminum adjuvant in infant vaccines. Their report concluded that there was no problem with aluminum in vaccines. Dr. Thomas Jefferson and colleagues stated that *"no evidence that aluminum salts cause any serious or long lasting adverse events.* ***Despite a lack of good-quality evidence we do not recommend that any further research on this topic is undertaken."*** Their questionable meta-analysis was done in 2004, which is well before aluminum adjuvants were substituted in numerous infant and childhood vaccines. Dr. Suzanne Humphries did an extensive research into vaccine safety and published a book titled *Dissolving Illusions,* in which she documented the historical framework of why vaccines have been an overhyped miserable failure. The toxicity of aluminum found in infant and adult vaccines far exceeds the toxicity of mercury. This is what Dr. Humphries had to say: *"Even if vaccines can prevent some infections, considering what is in them, there's no way they can improve overall health."* Dr. Kawahara had this to say about aluminum: ***"Whilst being environmentally abundant, aluminum is not essential for life. On the contrary, aluminum is a widely recognized neurotoxin that inhibits more than 200 biologically important functions and causes various adverse effects in plants, animals, and humans."*** According to Dr. Kawahara, aluminum promotes cellular death. Adverse effects of aluminum exposure are as follows:

1) DNA alteration, abnormal regulation of gene function, and gene expression interference
2) Damage in the cell membrane and myelin—the insulating layer around one's nerves—suffering and becoming dysfunctional
3) Binding to adenosine triphosphate (ATP); it affects one's energy metabolism.
4) Increased vascular endothelial aggressiveness and increased cardiovascular disease risk.
5) Coagulation of proteins, which can alter their function
6) Enhanced excitotoxicity in one's brain and increased brain inflammation

According to Dr. Sears, **"there is good evidence that large amounts of aluminum are harmful to humans."**

http://www.ask.drsears.com/topics/health-concerns/vaccines,

http://www.vaccinesummit.com

When adults and children are subjected to a vaccination regimen, especially established for kids by the health organizations, the aluminum compounds they contain accumulate not only at the injection site but travels to the brain and accumulates there. The aluminum enters neurons and glial cells.

What are glial cells? There are three types of glial cells in the central nervous system—astrocytes, oligodendrocytes, and microgial cells. The function of astrocytes is to maintain an appropriate chemical environment for neuronal signaling (synapsis). Function of the oligodendrocyte is restricted to the central nervous system, and to lay down a laminated lipid rich wrapping around some axons, called myelin. The function of the microgial cells is primarily as scavenger cells that remove cellular debris from sites of injury or normal cell turnover. Glial cells are involved in nearly every aspect of brain function, brain development, homeostasis, information processing, and neurological and psychiatric illness. These cells are also involved in synaptic transmission that implicates them in many aspects of learning, memory, and other types of information processing and in nervous dysfunction, including neurological and psychological disorders. Some glial cells mediate immune function to form electrical insulation on nerve axons (myelin).

http://www.ncbi.nlm.nih.gov/books/NBK10869/-http://www.thedoctorwillseeyounow.com/content/mind/art3792.html-109k

Aluminum is also linked to the so-called autoimmune/inflammatory syndrome (ASIA). An enormous body of research has begun to unravel how the environmental toxins, such as aluminum used in vaccines, can trigger an immune system reaction in susceptible individuals, and the potential exists for increased autoimmune diseases. So what is autoimmune disease? Autoimmune disease results when the body's system meant to attack foreign invaders turns instead to attack part of its own body it belongs to. The autoantibodies misidentify a particular part of the body and attack it. As an example, if they target the conductive neuron sheath (myelin), nerve impulses stop conducting properly, muscles go into a spasm and coordination fails, resulting in muscular sclerosis. Joint tissues are attacked, causing rheumatoid arthritis. There are numerous other effects.

A healthy immune system is tolerant to self-antigens. If this tolerance is disturbed, dysregulation results, and autoimmune disease follows. Vaccinations are one of the conditions that potentially disrupt the homeostasis in susceptible individuals. There are four types of individuals that are susceptible to autoimmune reaction due to vaccinations:

1) Individuals who have had a previous autoimmune reaction to a vaccine
2) Anyone with a medical history of autoimmunity
3) Patients with a history of allergic reactions
4) Anyone at high risk of developing autoimmune disease, including anyone that has a family history of autoimmunity, presence of autoantibodies, which are detectable by blood tests, low vitamin D, and those who smoke

An article provided from Philadelphia Children's Hospital listed the amount of aluminum present in some vaccines:

Pneumococcal vaccine = 0.125 mg/dose;
Diphtheria-tetanus-acellular pertussis vaccine—0.17 to 0.625 mg/dose
Haemophilus influenzae type b (Hib) vaccine—0.225 mg/dose
Hib/Hep 8 vaccine—225 mg/dose
Hepatitis A vaccine (hep A)—0.225 to 0.25 mg/dose in pediatric and 0.45 to 0.5 mg/dose for adults.
Hepatitis B vaccine (hep B)—0.225 mg/dose
HPV vaccine (Gardasil)—0.225 mg/dose.

http://www.vec.chop.edu./ . . . /vaccine-safety/
vaccine-ingredients/aluminum.html.

This article also lists quantities of aluminum in breast milk—0.04mg per liter; ponds, lakes, streams—0.1 mg/L; infant formula—0.225 mg/L; soy-based formula—0.46 to 0.83 mg/L; buffered aspirin—10 to 20 mg/L; antacid—104 to 208 mg/L, which is 104 to 208 ppm. Another source lists the HPV vaccine containing only 0.225 micrograms, which is considerably less than the mg amounts listed previously by the article from the Philadelphia Children's Hospital. This article also states, "Vaccines containing adjuvants are tested extensively in clinical trials before being licensed," **which is an outright lie.** One has to read the following information released by the FDA in 2002. FDA documentation from 2002 admits that **"routine toxicity studies in animals with vaccine ingredients such as aluminum adjuvants were never conducted because it was assumed that these ingredients are safe."**

Countries with heaviest vaccination schedules have higher autism rates compared to countries that do not vaccinate children with as many vaccines. There is compelling evidence that the human papillomavirus (HPV) raises the risk of brain autoimmune disorders, such as multiple sclerosis. Research also shows that repeated stimulation with the same antigen overcomes the genetic resistance to autoimmunity. Regular booster shots break one's tolerance to autoimmunity. A very interesting bit of information is that alum, which contains aluminum, was not used in vaccinations for these many years and would be refused registration on the basis of safety concerns. So as human beings, we receive aluminum from vaccines, from chemtrails, from food and water. Is it any wonder regarding our mental health? You be the judge, dear reader.

Dr. Suzanne Humphries made a very convincing presentation at the Vaccine Summit Conference against vaccination. Her presentation contained extensive peer-reviewed citations with references against the neurotoxin use of aluminum in vaccines.

http://www.vaccinationcouncil.org

http://www.healthimpactnews.com

http://www.vaccinesummit.com

Additional adjuvants that are used in vaccines are formaldehyde, human serum albumin, gelatin in the form of monosodium glutamate, antibiotics, and yeast proteins. How about the toxic effect of formaldehyde on the human body? (Toxic effects are covered under aspartame chapter.) The reason given for their use is that these adjuvants help the vaccine to work better. Human albumin helps to stabilize live viruses. Formaldehyde, antibiotics, egg protein, and yeast proteins are all residual amounts left over from the way the vaccines are made. The presence of formaldehyde in some childhood vaccines, included in the flu shots, polio vaccine, and DTaP vaccine, is to eliminate the harmful effects of these bacterial toxins and makes the viruses unable to reproduce themselves. The claim is made that it is in very small amounts. How about the amount in food or baby formula in the form of aspartame and fluoride in the water used in numerous liquids, which also exerts a toxic effect on the brain? I guess it is all additives to a potent toxic level in brain or some other organ or gland to a point exhibited in adverse health effects.

Some interesting events have taken place recently regarding vaccines. Dr. William Thompson, former high-level scientist at the CDC, broke silence on fraud at the CDC that led to crucial data linking vaccines to autism, being censored and withheld. Dr. Thompson, jointly with Dr. Brian Hooker and Dr. Andrew Wakefield, produced a video explaining how the CDC altered study data to make it seem as though vaccines and vaccine ingredients and autism are unrelated, when in fact they are **deeply connected**. Since increasing reports are coming in on YouTube by mothers from autistic children. As an example, one mother of eight children reported the following. "Out of my vaccinated children, three have autism, one has ADHD, one has severe mood swings, and one has severe language problems. My two unvaccinated children have none of their sibling's disorders."

http://www.youtube.com

http://www.vimeo.com

http://science.naturalnews.com

http://www.naturalnews.com/046750_autistic_children_vaccines_CDC.html#ixzz3CU78daax

How about the latest news regarding the outbreak of the mystery virus EV-D68 among vaccinated children? Hundreds of children are hospitalized across eight mid-west states. CNN reported that "about 475 children were recently treated at Children's Mercy Hospital in Kansas City, and at least 60 had to receive intensive hospitalization."

"It's worse in terms of scope of critically ill children who require intensive care, I would call it unprecedented. I've practiced for 30 years in pediatrics, and I've never seen anything quite like this," said Dr. Mary Ann Jackson, the hospital's division director for infectious diseases. The interesting part is that the outbreak is by children who have been vaccinated with MMR vaccines, influenza vaccines, polio vaccines, and many others are the same children who are now struck with EV-D68. How was this determined? The news media is usually quick to blame any outbreak on unvaccinated individuals. Any inconvenient fact about vaccines that the pro-vaccine media doesn't want the public to know, it will withhold that fact from the stories. This was made clear recently when the media wide blackout of the CDC whistleblower story by Dr. William Thompson, who openly admitted to committing scientific fraud at the CDC in order to hide the link between MMR vaccines and autism.

http://www.naturalnews.com/046810_pandemic_outbreak_EV-D68_vaccines.html#ixzz3CpO7Xpss

Another very interesting story came from the health minister of African country Kenya regarding the tetanus vaccine. But first, here is a little background history regarding this vaccine. These are some excerpts as presented by Mike Adams, publisher of Natural News. In the early 1990s, the World Health Organization (WHO) was involved in a vaccination campaign project against tetanus in a number of countries, among which was Nicaragua, Mexico, and Philippines.

WHO has been extensively involved in development of an anti-fertility vaccine that utilizes the hCG hormones tied to the tetanus toxoid and other carriers. According to the report by Natural News, some of the organizations tied to this development of antifertility vaccine (AFV) that use tetanus as carrier, have been the following:

UNFPA, UN Development Programme 9UNDP0, the World Bank, the Population Council, the Rockefeller Foundation, the All India Institute of Medical Sciences, Universities Uppsala, Helsinki, and Ohio State, among others.

The Comite Pro Vida de Mexico obtained several vials of the vaccine and had them analyzed. Some of the vials were found to contain human chorionic gonadotropin (hCG), which is a naturally occurring hormone essential for maintaining a pregnancy. The hCG hormone alerts the woman's body that she is pregnant, causing a release of a number of hormones to prepare the uterus for implanting the fertilized egg. This is the hormone a woman tests for to determine if she is pregnant.

When this hCG is introduced into the body with the tetanus toxoid carrier, antibodies are formed not only against the tetanus toxoid but also against hCG. The antibodies then attack subsequent pregnancies by killing the naturally occurring hCG that are to sustain a pregnancy. With sufficient anti-hCG antibodies in a woman's system, the woman is rendered incapable of maintaining a pregnancy.

The protocol that was followed for vaccination was to only vaccinate women between the ages of fifteen to forty-five, and it calls for multiple vaccinations, a total of five injections over a two-year time frame. The Philippines reported first about the hCG presence in the vaccinations. WHO and the Department of Health of the Philippines denied the content of hCG in the vaccines. A follow-up of additional vial tests confirmed that the vaccines were laced with hCG. From outright denial, the story shifted to insignificant quantity. A blood test of twenty-six women out of thirty that were tested confirmed the presence of high levels of hCG in their blood. There is no known way for the vaccinated women to have hCG antibodies in their blood unless it was artificially introduced into their bodies.

While the vaccine controversy was raging, it was reported that hCG had *not* been licensed for sale or distribution or registered with the Philippine Bureau of Food and Drugs, which was required by law. One of the manufacturing companies was Connaught Laboratories Ltd. of Canada, which during the mid-'80s was found to be knowingly distributing vials of AIDS-contaminated blood product.

http://www.naturalnews.com/047571_vaccines_sterilization_genocide.html?utm_content=buffer55aa4&utm_medium=social&utm_source=facebook.com&utm_campaign=buffer#ixzz3loVtthYd

Fast forward to year 2014. Kenya's Catholic bishops are charging two United Nations organizations with sterilizing millions of girls and women under cover of an antitetanus inoculation program sponsored by Kenyan government. Dr. Muhame Ngare reported the following: "We sent six samples from around Kenya to laboratories in South Africa. They tested positive for the hCG antigen, that were all laced with hCG antigen." Later, he released the following statement: "This proved right our worst fears; that this WHO campaign **is not about eradicating neonatal tetanus but a well coordinated forceful population control, mass sterilization exercise using a proven fertility regulating vaccine."** Of course, the health minister, James Macharla, and head of ministry's immunization branch, Dr. Collins Taby, told the Kenyan nation that 'there is no other additive in the vaccine other than the tetanus antigen. Women who were vaccinated in October 30, 2013 and March this year are expecting." To which Dr. Ngare responded, "We knew that the last time this vaccination with five injections has been used in Mexico in 1993 and Nicaragua and the Philippines in 1994. It didn't cause miscarriages till three years later, therefore, the counterclaim that women who got the vaccination recently and then got pregnant are meaningless."

Source: http://www.lifesitenews.com/news/a-mass-sterilization-exercise-kenuan-doctors-anti-fertility-agent-in-

Another very interesting story these past few months in 2014. The outbreak of Ebola in West African nations. The very strange information is that CDC has a patent number on this pathogen issued in 2004. The big question then comes to mind how did this virus surface in four of the West African countries at the same time? Of course WHO, CDC, and some international health organizations scrambled to control this outbreak. CDC was hard at work to come up with a vaccine. The cry went out to find a cure for the Ebola virus. The most interesting situation arose when a Natural Solutions Foundation (NSF) shipped two hundred bottles of nanosilver 10 ppm to Sierra Leon, as a potential cure against Ebola. The reason for sending nanosilver was that the Department of Defense's Defense Threat Reduction Agency and the US Strategic Command Center for Combating Weapons of Mass Destruction revealed that antimicrobial silver solutions have proven to be beneficial in fighting Ebola and other forms of hemorrhagic fever. When WHO learned of this, the minister of health of Sierra Leon was instructed that if he uses it, the country will lose all their financial assistance. The colloidal silver was returned back to the US.

http://www.thesilveredge.com

http://www.npr.org

http://science.naturalnews.com

http://www.naturalnews.com/047101_Ebola_colloidal_silver_government_seizure.html#3TXLyxC4f

People are dying, but bureaucrats are playing politics. The reason given was to develop a vaccine against Ebola or . . . ?

Another very interesting story regarding vaccination was the outbreak of whooping cough that occurred in Park City, Utah. The most perturbing news was that the outbreak occurred among children who were vaccinated against whooping cough. Pertussis (whooping cough) is highly contagious, but the problem for Park City officials is that all the infected children were up-to-date with their vaccinations. The parents of these children are trying to get an answer as to why vaccinated kids came down with the disease.

The truth is that the medical community has known about the pertussis vaccine dangers. A California study after an outbreak of pertussis showed that 81% of pertussis cases occurred in children that were vaccinated for it, but only 8% of the cases that were not vaccinated for whooping cough. The potential conclusion that can be drawn is that the vaccine for whooping cough (pertussis) is not effective but is causing more cases of whooping cough sickness in children. Yet the medical profession is still advocating more vaccination.

http://www.naturalhealth365.com/whooping-cough-outbreak-vaccine-dangers-1374.html#sthash.XxmBzUrT.dpuf http://www.smartvax.com/index.php?option=com_content&view=article&id=112

http://www.science.slashdot.org/story/12/04/19/154206/in-calif-study-most-kids-with-whooping-cough-were-fully-vaccinated

HPV Gardasil Vaccines

One of the more controversial vaccines released is HPV vaccine under the name of Gardasil. The reason for its manufacture was to vaccinate young girls at about age eleven or twelve years old to prevent the human papilloma virus (HPV). There are about a one hundred different types, but about forty of those are sexually transmitted, and fifteen of those are associated with cervical cancers and genital warts in women and men. Left unidentified and untreated for a long time, these conditions can develop into vaginal, vulvar, penile, anal, and oropharyngeal cancers. These infections can cause minor skin infections and common warts on hands and feet. Certain types of HPV infection can develop into cervical cancer. This is usually if the virus has not been identified and left untreated for a lengthy time that results in the formation of abnormal cervical cells that can turn into cancer. Usually this is uncovered by the Pap smear test which identifies cervical changes. This test is far more effective since it identifies cervical abnormalities resulting in time to find it and treat the condition. The death rate due to cervical cancer in the US is about 3 per 100,000. It was estimated that in 2011 about 12,000 women were diagnosed with cervical cancer and 4,000 died. The cancer rates are even lower in some European countries. Reason for such low mortality rate is that women living in developed countries have a very good immune system that is strong enough to naturally clear HPV infection within two years. This is accomplished in 90% of all cases.

The vaccine contains only four HPV proteins. These are HPV 6 L1, HPV 11 L, HPV 16 L, and HPV 18 L. There are over forty to one hundred different HPV viruses, so the vaccine contains only four. It is also known that Gardasil contains genetically engineered virus-like protein particles as well as aluminum. According to the manufacturer product information insert, the vaccine has not been evaluated for the potential to cause cancer or if it is toxic.

Here are the three main HPV vaccines, and the new revised recommendations regarding vaccination: **What are the recommendations?** "9vHPV, 4vHPV or 2vHPV can be used for routine vaccination of females aged 11 or 12 years and females through age 26 years who have not been vaccinated previously or who have not completed the 3-dose series. 9vHPV or 4vHPV can be used for routine vaccination of males aged 11 or 12 years and males through age 21 years who have not been vaccinated previously or who have not completed the 3-dose series. ACIP recommends either 9vHPV or 4vHPV vaccination for men who have sex with men and immunocompromised persons (including those with HIV infection) through age 26 years if not vaccinated previously."

TABLE 1. Characteristics of the three human papillomavirus (HPV) vaccines licensed for use in the United States:

Characteristic	**Bivalent (2vHPV)***	**Quadrivalent (4vHPV)†**	**9-valent (9vHPV)§**
Brand name	Cervarix	Gardasil	Gardasil-9
VLPs	16, 18	6, 11, 16, 18	6, 11, 16, 18, 31, 33, 45, 52, 58
Manufacturer	GlaxoSmithKline	Merck and Co., Inc.	Merck and Co., Inc.
Manufacturing	*Trichoplusia ni* insect cell line infected with L1 encoding recombinant baculovirus	*Saccharomyces cerevisiae* (Baker's yeast), expressing L1	*Saccharomyces cerevisiae* (Baker's yeast), expressing L1

Adjuvant	500 µg aluminum hydroxide, 50 µg 3-O-desacyl-4' monophosphoryl lipid A	225 µg amorphous aluminum hydroxyphosphate sulfate	500 µg amorphous aluminum hydroxyphosphate sulfate
Volume per dose	0.5 ml	0.5 ml	0.5 ml
Administration	Intramuscular	Intramuscular	Intramuscular

Abbreviation: L1 = the HPV major capsid protein; VLPs = virus-like particles.
* Only licensed for use in females in the United States. Package insert available at http://www.fda.gov/downloads/BiologicsBloodVaccines/Vaccines/ApprovedProducts/UCM186981.pdf, http://www.fda.gov/downloads/BiologicsBloodVaccines/Vaccines/ApprovedProducts/UCM186981.pdfhttp://www.cdc.gov/Other/disclaimer.html, http://www.cdc.gov/Other/disclaimer.html.
† Package insert available at http://www.fda.gov/downloads/BiologicsBloodVaccines/Vaccines/ApprovedProducts/UCM111263.pdf http://www.fda.gov/downloads/BiologicsBloodVaccines/Vaccines/ApprovedProducts/UCM111263.pdf, http://www.cdc.gov/Other/disclaimer.htmlhttp://www.cdc.gov/Other/disclaimer.html.
§ Package insert available at http://www.fda.gov/downloads/BiologicsBloodVaccines/Vaccines/ApprovedProducts/UCM426457.pdf http://www.fda.gov/downloads/BiologicsBloodVaccines/Vaccines/ApprovedProducts/UCM426457.pdf, http://www.cdc.gov/Other/disclaimer.htmlhttp://www.cdc.gov/Other/disclaimer.html.

New ACIP Recommendation for Human Papillomavirus Vaccination

Effective March 27, 2015

- 9vHPV, 4vHPV, or 2vHPV for routine vaccination of females eleven or twelve years of age and females through twenty-six years of age who have not been vaccinated previously or who have not completed the three-dose series.

- 9vHPV or 4vHPV for routine vaccination of males 11 or 12 years of age and males through twenty-one years of age who have not been vaccinated previously or who have not completed the three-dose series.
- 9vHPV or 4vHPV vaccination for men who have sex with men and immunocompromised men (including those with HIV infection) through age twenty-six years if not vaccinated previously.

Syndicated and print schedules do not yet reflect this change. This recommendation will be incorporated in the immunization schedules that will be published in February 2016.

What is currently recommended?

The Advisory Committee on Immunization Practices (ACIP) "recommends routine HPV vaccination at age 11 or 12 years. The vaccination series can be started beginning at age 9 years. Vaccination is also recommended for females aged 13 through 26 years and for males aged 13 through 21 years who have not been vaccinated previously or who have not completed the 3-dose series. Males aged 22 through 26 years may be vaccinated. ACIP recommends vaccination of men who have sex with men and immunocompromised persons through age 26 years if not vaccinated previously."

See MMWR for complete vaccine recommendations.

Merck, the manufacturer of Gardasil, studied the vaccine in fewer than 1,200 girls under sixteen years old, prior to its release to the market under a fast track road to licensure. Most of the serious side effects, including death, that occurred during this time frame in clinical trials and post-marketing surveillance were written off as a "coincidence" by the company's researchers and government officials.

Reprinted from CDC ACIP (Advisory Committee on Immunization) regarding HPV vaccine safety studies:

> Safety has been evaluated in approximately 15,000 subjects in the 9vHPV clinical develop-

> ment program; approximately 13,000 subjects in six studies were included in the initial application submitted to FDA (*2*). The vaccine was well-tolerated, and most adverse events were injection site-related pain, swelling, and erythema that were mild to moderate in intensity. The safety profiles were similar in 4vHPV and 9vHPV vaccines. Among females aged 9 through 26 years, 9vHPV recipients had more injection-site adverse events, including swelling (40.3% in the 9vHPV group compared with 29.1% in the 4vHPV group) and erythema (34.0% in the 9vHPV group compared with 25.8% in the 4vHPV group). Males had fewer injection site adverse events. In males aged 9 through 15 years, injection site swelling and erythema in 9vHPV recipients occurred in 26.9% and 24.9%, respectively. Rates of injection-site swelling and erythema both increased following each successive dose of 9vHPV.

There is no mention of any highly adverse health problems. The only recommendation stated is this: "Adverse events occurring after administration of any vaccine should be reported to VAERS." Additional information about VAERS is available by telephone (1–800–822–7967) or online at http://vaers.hhs.govhttp://www.cdc.gov/Other/disclaimer.htmlhttp://www.cdc.gov/Other/disclaimer.html.

On the National Vaccine Information Center's (NVIC) website, one can access and read about description of women and girls who have suffered serious health deterioration after Gardasil shots. In some cases, death occurred within a short time. Following, are three examples from an article from Dr. Mercola, titled "213 Women Who Took Gardasil Suffered Permanent Disability."

* Christina Tarsell, a twenty-one-year-old college student majoring in studio arts at Bard College, died suddenly

and without explanation shortly after receiving the third Gardasil shot in June 2008.

* Megan, a twenty-year-old college student died suddenly without explanation, about one month after receiving her third Gardasil shot. No cause of death was found.
* Ashley, a sixteen-year-old, became chronically ill after receiving Gardasil and now suffers regular life-threatening episodes of seizure-like activity, difficulty in breathing, back spasm, paralysis, dehydration, memory loss, and tremors.

Reports from Australia have come that a woman by the name of Naomi Snell has filed a class-action lawsuit against drug manufacturer Merck due to suffering neurological and autoimmune problems after injections with the HPV vaccine Gardasil.

Another case is that of Gabi Swank, a fifteen-year-old honor student, who decided to get the Gardasil vaccine after seeing a Gardasil ad on TV. There was no warning about possible side effects on the series of three shots. She suffered two strokes and experienced partial paralysis. She had to use a wheelchair on numerous occasions while attending high school.

Here is what Dr. Ian Sutton, a neurologist, had to say regarding the patients he saw: *"We report five patients who presented with multifocal or atypical demyelinating syndromes within 21 days of immunization with the quadrivalent human papilloma virus (HPV) vaccine, Gardasil. Although the target population for vaccination, young females, has an inherently high risk for MS, the temporal association with demyelinating events in these cases may be explained by the potent immuno-stimulatory properties of HPV virus-like particles which comprise the vaccine."*

Reported are increasing numbers of Gardasil Multiple-Sclerosis-like symptoms, **neurological, seizures, paralysis and speech problems by girls and women following Gardasil vaccination.** Judicial Watch reported, "Between May 2009 and September 2010, 16 deaths after Gardasil vaccination were reported. For that timeframe, there were also 769 reports of 'serious' Gardasil adverse reactions, including 213 cases of permanent disability and 25 diagnosed cases of Guillain-Barre Syndrome."

Guillain-Barre syndrome is a rare disorder in which the body's own immune system mistakes parts of the peripheral nerves for an infection. It sends out antibodies that attack those nerves. The result most commonly identified are weakness and numbness that starts at the tips of fingers and toes and spreads inward toward the body. Sometimes this weakness is so severe that the patient cannot breathe on their own.

Additional potential side effects listed are pain, swelling, itching, bruising, redness at the injection site, headache, fever, nausea, dizziness, vomiting, and fainting. The patient therefore is requested to stay an additional fifteen minutes for observation after injection. More severe symptoms are difficulty in breathing, wheezing, hives, pale skin, and fast heartbeat. It is also not recommended for anyone who has had an allergic reaction to yeast, to a prior dose, or to any Gardasil vaccine components. It is also not recommended who is currently taking immunosuppressant therapy or is immunocompromised, or someone who has a moderate ailment with fever greater than 100 degrees F. Women who are currently pregnant or planning pregnancy during the course of treatment. Also someone who has a bleeding disorder in which injections are contraindicated. One should also understand that Gardasil vaccine will not protect against other sexually transmitted diseases, such as chlamydia, gonorrhea, herpes, HIV, syphilis, and trichomoniasis. If after the first Gardasil injection one develops life-threatening allergic reaction, do not receive a booster shot.

One should also inform the health professional prior to Gardasil injection regarding all other vaccines received recently. The doctor should be informed about any recently used drugs or treatments that can weaken the immune system. The information should include all prescription and over-the-counter medications. The list should include vitamins, supplements, minerals, and herbal products. This is quite a list of information to be given to a health professional prior to starting Gardasil inoculation.

On December 10, 2014, the FDA approved a second generation of Gardasil-9 vaccine. The claim is that this vaccine is improved over the initial vaccine by increasing the number of HPV proteins, from initial four to nine. FDA approval letter states that "the action to approve was done without the consent of the Vaccines and Related Biological Products Advisory Committee (VRBPAC)." This committee is responsible for reviewing and evaluating data concerning safety, effectiveness, and appropriate use of vaccines and related biological products. The reason stated by the director of Office of Vaccines Research and Review is: "We did not refer your application to the Vaccines and Related Biological Products Advisory Committee because our review of information submitted in your BLA, including the clinical study design and trial results, did not raise concerns or controversial issues which would have benefited from an advisory committee discussion." **Are you kidding us? After numerous neurological and even deaths resulting from the initial vaccine, this did not warrant a discussion? Better yet, it should have been denied. Where are these so-called guardians of our health?** According to the mission statement, FDA is **"responsible for protecting the public health by assuring safety, efficacy and security of human and veterinary drugs, biological products, medical devices, our nation's food supply, cosmetics, and products that emit radiation."**

This does not bode well for our so-called guardians of our health. This decision is very disturbing due to the numerous reports worldwide regarding the safety and effectiveness of the vaccine. Comparison from the package inserts for Gardasil and Gardasil-9—here are the following differences:

Gardasil	Ingredient	Gardasil-9
225 mcg	Aluminum	500 mcg
0.56 mcg	Sodium Chloride	9.56 mcg
0.78 mcg	L-Histidine	0.78 mcg
50 mcg	Polysorbate	50 mcg
35 mcg	Sodium Borate	35 mcg
7 mcg	Yeast Protein	7 mcg
20 mcg	HPV 6 L1	30 mcg
40 mcg	HPV 13 L1	40 mcg
40 mcg	HPV 16 L1	60 mcg
20 mcg	HPV 18 L1	40 mcg
	HPV 31 L1	20 mcg
	HPV 33 L1	20 mcg
	HPV 45 L1	20 mcg
	HPV 52 L1	20 mcg
	HPV 58 L1	20 mcg

As can be observed, the aluminum content has been more than doubled, a known neurotoxin. If one looks up aluminum neurotoxicity on PubMed there are well over a thousand peer-reviewed public scientific papers on that subject. Also the amount of three of the HPV L1 contents was also increased.

Following is a comparison of initial Gardasil vaccine to Gardasil-9 on adverse effects that the FDA defined as death, threat to life, hospitalization, disability or permanent damage, congenital abnormality/birth defects, or the requirement to intervene to prevent permanent impairment.

Serious adverse events comparison between the two vaccines

Num. received inoculation	Type of vaccine	% Serious AE	Num. serious AE
13,236	Gardasil-9	2.30%	305
7,378	Gardasil	2.50%	185

What does that mean? Cervical cancer is reported as a number per 100,000. Now based on Gardasil-9 numbers, you have about 2,300 instances of serious adverse effects. Is that worth the risk in getting vaccinated by the vaccine? The other problem that HPV vaccine does is the onset of autoimmune disorders. That is when the body's immune system attacks and destroys healthy body tissue by mistake. During Merck's initial trial, the medical conditions potentially indicative of autoimmune disorders were as follows:

Num. receiving shot	Type of vaccine	% Autoimmune Disorder	Num. AD
13,236	Gardasil	2.40%	321
7,378	Gardasil-9	3.30%	240

Therefore, in addition of serious adverse effects, now an additional number of 2,300 people may have autoimmune problems. According to Dr. Lucija Tomljenovic, who did an extensive study on Gardasil vaccine reports from numerous countries, "*The presence of cross-reactive anti-HPV antibodies in the blood of vaccinated girls will increase their risk for developing immune-mediated nervous system disorders, which incidentally appear to be the most commonly reported worldwide, following Gardasil. We have done the analysis on adverse events reported following HPV vaccination. It was publishedt in the Annals of Medicine. We took vaccines safety databases from various countries, and then we rated the adverse effects reported based on organ system. We found the most commonly reported adverse events following HPV vaccination are nervous system disorders of immune origin."*

http://vaccines.mercola.com/sites/vaccines/download-nvic-posters.aspx

Since that time, the Japanese Ministry of Health has decided to cease inoculations of Gardasil in Japan of young girls.

Another statement in the Gardasil-9 package insert indicated that 1,028 women were injected with the vaccine during the clinical trial along with 991 women who received the original Gardasil vaccine, suffered adverse effects. A total of 313 women either lost their babies to spontaneous abortion or late fetal death or gave birth to children with congenital anomalies. This translates into a 27.4% rate. But Merck states, **"The proportion of adverse outcomes observed were consistent with pregnancy outcomes observed in the general population."** According to CDC's latest publication on fetal mortality, the rate of spontaneous abortions and fetal deaths in the US is 6.05/1,000 or 0.605%, which is much less than the 27.4%.

For a recently informative discussion/reporting on the Gardasil vaccine and any other vaccinations one should go to this website. This website has a plethora of reports regarding vaccinations. Another website that is also very informative regarding vaccinations is PEERS at http://www.WantToKnow.com, and access their index listed under vaccinations

http:://www.healthimpactnews.com/2014/stronger-more-toxic-gardasil

http://www.truthaboutgardasil.org

http://www.webmed.com/vaccines/features/should-your-child-get-hpv-vaccine

http://www.drugs.com/sfx/gardasil-side-effects

http://www.articles.mercola.com/sites/articles/archive/2012/01/24/hpv-vaccine-victim-sues-merck

http://www.naturalnews.com/047024_HPV_vaccine_Gardasil_GeoffreySwain#jxzz3Uj2fCDrB

In the Western world, cervical cancer is a rare disease with mortality rates that are several times lower than the rate of reported serious adverse reactions (including deaths) from HPV vaccination. Future vaccination policies should adhere more rigorously to evidence-based medicine and ethical guidelines for informed consent.

One of the independent laboratory analysis run by Dr. Lee discovered that the Gardasil vaccine contained viral DNA, which was contrary to Merck's statement, claiming "that Gardasil contained no viral DNA." The viral DNA fragments discovered by Dr. Lee were firmly attached to Merck's proprietary aluminum adjuvant in all the samples. What Dr. Lee discovered was that these fragments had also adopted a non-B DNA conformation, a novel new chemical compound of unknown toxicity. According to Dr. Lee, "This is very interesting since this is the first time I came across this new identified chemical compound presence in any vaccine." My search of the Internet discovered this article by *Regina Z. Cer, Duncan E. Donohue, Uma S. Mudunuri, Nuri A. Temiz, Michael A. Loss, Nathan J. Starner, Goran N. Halusa, Natalia Volfovsky, Ming Yi, Brian T. Luke, Albino Bacolla, Jack R. Collins, and Robert M. Stephens* (Nucl. Acids Res. (2013) 41 (D1): D94-D100. doi: 10.1093/nar/gks955 \l " Non-B DB v2.0: a database of predicted non-B DNA-forming motifs and its associated tools).

The FDA was quick to confirm this, but stated it did not pose any health risks. It is reproduced here: **non-B DB** "DNA exists in many possible conformations that include the A-DNA, B-DNA, and Z-DNA forms; of these, B-DNA is the most common form found in cells. The DNAs that do not fall into a right-handed Watson-Crick double-helix are known as non-B DNAs and comprise cruciform, triplex, slipped (hairpin) structures, tetraplex (G-quadruplex), left-handed Z-DNA, and others. Several recent publications have provided significant evidence that non-B DNA structures may play a role in DNA instability and mutagenesis, leading to both DNA rearrangements and increased mutational rates, which are hallmark of cancer." And "repeating sequences on their non-B DNA conformation, causes gross genomic rearrangements (translocation, deletions, inversions, and duplications). These rearrangements are the genetic basis for human genome diseases, including polycystic kidney disease, adrenoleukodystrophy, follicular lymphoma, and spermatogenic failure. At least 70% of diseases fall into this category. To tease apart the contributions of DNA sequence versus non-B DNA conformation adopted by the sequences, a family of experiments was developed that rigorously demonstrated that the non-B DNA confrontations are the culprits regarding mutagenesis."

http://www.ncbi.nim.gov/pmc/articles/PMC266547/

According to *Wikipedia*'s definition of mutagenesis, "It is a process by which the genetic information of an organism is changed in a stable manner, resulting in a mutation. It may occur spontaneously in nature, or as a result of exposure to mutagens. It can also be achieved experimentally using laboratory procedures. In nature, mutagenesis can lead to cancer and various heritable diseases, but it is also a driving force of evolution. Mutagenesis is a science that was developed based on work done by Hermann Muller, Charlotte Auerbach and J. M. Robson in the first half of the 20th century. Mammalian nuclear DNA may sustain more than 60,000 damages per cell per day, as listed with references in DNA damage (naturally occurring). If left uncorrected, these adducts, after misreplication past the damaged sites, can give rise to mutations. In nature, the mutations that arise may be beneficial or deleterious, it is the driving force of evolution, an organism may acquire new traits through genetic mutation, but mutation may also result in impaired function of the genes, and in severe cases, causing the death of the organism. In the laboratory, however, mutagenesis is a useful technique for generating mutations that allows the functions of genes and gene products to be examined in detail, producing proteins with improved characteristics or novel function, as well as mutant strains with useful properties. Initially the ability of radiation and chemical mutagens to cause mutation was exploited to generate random mutations, but later techniques were developed to introduce specific mutations."

For a really highly informative discussion/reporting on the Gardasil vaccines and any other vaccinations, please go to http://www.SaneVac.com, they have multitude of reports regarding vaccinations. Other good useful information is on PEERS at http://www.WantToKnow, and access their index listing for vaccinations.

http://www.healthimpactnews.com/2014/stronger-more-toxic-gardasil

http://www.truthaboutgardasil.org

http://www.webmed.com/vaccines/features/should-your-child-get-hpv-vaccine

http://www.drugs.com/sfx/gardasil-side-effects

http://www.articles.mercola.com/sites/articles/archive/2021/01/24/hpv-vaccine-victim-sues-merck

http://www.naturalnews.com/047024_HPV_vaccine_Gardasil_GeoffreySwain.html#jxzz3Uj2fCDrB

In conclusion, Dr. Sin Hang Lee's statement best summarizes HPV vaccine: "**HPV vaccination is unnecessary and potentially dangerous to some recipients. This is the first vaccine invented by the government, patented by the government, approved by the government, regulated by the government and promoted by the government to prevent an already preventable disease (cervical cancer) 30 years down the road based on using a poorly demarcated, self-reversible surrogate end-point (CIN2/CIN3) for evaluation of vaccine efficacy, A BIG SCIENTIFIC FRAUD. There is no cervical cancer epidemic in any developed countries.**"

Evidence continues to grow that the vaccine harms children. This is a pharmaceutical product that a large number of experts characterize it as **overly expensive, ineffective, and potentially deadly.** There is a mounting evidence of its highly harmful health effects and even death. The physicians, pharmaceutical companies and numerous TV ads should not be believed. Dr. Bernard Dalbergue, a former pharmaceutical physician for the vaccine's manufacturer, Merck, stated, ***"The evidence will add up to prove that this vaccine, technical and scientific feat that it may be, has absolutely no effect on cervical cancer and that all the very many adverse effects which destroy lives and even kill, serve no other purpose than to generate profit for the manufacturers."***

Some of the complications associated with Gardasil vaccine, beside the **pain and fever, include such debilitating and deadly complications as paralysis of the lower limbs, vaccine induced multiple sclerosis, cases of Guillain-Barre syndrome, vaccine-in-**

duced encephalitis, and sudden death. This is all contrary to what CDC, health officials, main news media and the pharmaceutical companies that they disseminate in their informative advertising.

http://www.collective-evolution.com/2015/01/25/mercks-former-doctor-gardasil-to-become-the-greatest-medical-scandal-of-all-time

http://sanevax.org/gardasil-international-scandal

http://www.naturalhealth365.com/side-effects-of-gardasil-shot-vaccine-dangers-1349.html#sthash.9sVc7G4s.dpuf

The latest news that has surfaced comes from Spain. It joins a growing list of countries to file criminal complaints against Gardasil manufacturers. The filed complaint states that Merck Laboratories *"failed to use inert placebo during clinical trials, thereby manipulating data and marketing Gardasil under false pretenses. Despite complaints of several young women with similar new medical conditions after Gardasil injections, the Spanish health authorities ignored calls for a moratorium on the use of Gardasil until the safety issues were resolved. Both regional and national health authorities made no attempt to verify the accuracy and safety data Merck submitted to gain approval for the widespread administration of Gardasil as a cancer preventative; nor did they make any attempt to inform the public that an already proven safe and effective means of controlling cervical cancer was already in existence . . . The attitude of the Merck pharmaceutical company and Spanish health authorities before, during and after the administration of Gardasil shows they care nothing about the risk to which medical consumers expose themselves whenever Gardasil is used."* According to Attorney Don Manuel Saez Ochoa, *"(claiming) a possible exemption possible arguing that they did not know at the time of processing, the dangers of the vaccine (Gardasil) is laughable . . . Frankly this attitude seems clearly malicious and constitutes the offense of injury as per Article 149.1 of the Criminal Code that states: To cause another, or by means or process, the loss or worthlessness of an organ or principal member, or a sense, impotence, sterility, severe deformity, or severe somatic or mental illness, shall be punished with imprisoment of six to twelve years."*

To be more specific, the company Merck—Sanofli Pasteur, Spain's National and Regional health authorities are charged with the following:

- fraudulent marketing and/or administration of an inadequately tested vaccine
- failure to inform the public about the potential risks of using Gardasil
- clear infringement of the right to informed consent
- ignoring new medical conditions in those who used Gardasil despite the similarity of their symptoms and the relatively short period of time between vaccine administration and the onset of symptoms
- ignoring established and new scientific evidence illustrating the potential harmful effects of Gardasil ingredients and manufacturing methods
- callous disregard for those suffering new medical conditions post-Gardasil
- failure to inform the public that HPV infections are simply one of the risk factors involved in the development of cervical cancer
- failure to inform the public that 90% of all HPV infections clear on their own without medical intervention
- failure to inform the public about alternative methods of controlling cervical cancer
- criminal liability for the injuries resulting from the administration of Gardasil

http://www.naturalnews.com/049256_Gardasil_HPV_vaccine_Spain.html#ixzz3WqLy3J14,
http://www.healthimpactnews.com/ . . . /-191k

Japan has issued a suspension of the government's recommendation to get vaccinated against HPV in June 2013. The government and health authorities organized a symposium on the HPV vaccine. A very important testimony was delivered by a doctor who treated twenty cases of MS following Gardasil injection. The pharmaceutical representatives tried to discredit the report that these side effects are psychogenic. That is their favorite explanation. Problem is how does a psychogenic disorder cause MS lesions in a person's brain? How about a healthy girl prior to vaccination? This was reported to Dr. Tomljenovic, who compiled some interesting statistics regarding Gardasil side effects. She stated, "*All these problems started in temporal association with the vaccine. Just our precautionary principle, you would think that they would have the common sense to at least half the use of the vaccine until more research is done. But no, they just want to force it, and they parrot that it is safe. They do not have any proof of safety other than manipulated research.*"

The manufacturers of Gardasil (Merck) and Cervarix (GlaxoSmithKline) tried to have vaccines reinstated in Japan. The attempts were futile. Japan did not reinstate vaccinations with the HPV vaccines.

Very good additional reading regarding Gardasil vaccine is on the following links:

Dr. Russell Blaylock exposes criminal fraud of Gardasil, HPV vaccinations: http://www.naturalnews.com/036874_Dr_Russell_Blaylock_Gardasil_HPV_vaccines.html#ixzz3WY0x2DRP

http://tv.naturalnews.com/v.asp?v=4D703FEAA094BED0DB02BEDC4507765C

http://www.russellblaylockmd.com/

http://tv.naturalnews.com/v.asp?v=4D703FEAA094BED0DB02BEDC4507765C

http://www.naturalnews.com/Gardasil.html

http://www.naturalnews.com/036874_Dr_Russell_Blaylock_Gardasil_HPV_vaccines.html#ixzz3WY1NvNGD

More news on Gardasil

Gardasil vaccination: Evaluating the risks versus benefits

Dr. Russell Blaylock exposes criminal fraud of Gardasil, HPV vaccinations

Blood donations from people vaccinated against HPV may be harmful to recipients

Gone after Gardasil: Acceptable collateral damage?

Most Gardasil 'safety' data discovered to be fraudulent

Parents blame their daughter's mysterious death on Gardasil vaccine, send in brain tissue for evaluation

HPV vaccines: Gardasil becomes a market dud in wake of informed backlash

http://www.naturalnews.com/049256_Gardasil_HPV_vaccine_Spain.html#ixzz3WY2r3zrd

Gone after Gardasil: Acceptable collateral damage?

References:

1. http://www3.cfo.com
2. http://www.ncbi.nlm.nih.gov/pubmed/21944226 (see table 3, The ATHENA human papillomavirus study: design, methods, and baseline results.) http://www.scribd.com/fullscreen/80272698
3. http://sanevax.org/gone-after-gardasil-jessica-new-york/
4. http://sanevax.org/gone-after-gardasil-christina-maryland/
5. http://sanevax.org/gone-after-gardasil-annabelle-canada/
6. http://sanevax.org/gone-after-gardasil-jasmine-new-zealand/
7. http://sanevax.org/gone-after-gardasil-megan-new-mexico/

http://www.naturalnews.com/034890_Gardasil_collateral_damage_fatalities.html#ixzz3WY4If3L1

Blood donations from people vaccinated against HPV may be harmful to recipients

http://www.naturalnews.com/042227_HPV_vaccines_blood_donations_adverse_reactions.html#ixzz3WY4gpco7

http://sanevax.org
http://vactruth.com
http://www.naturalnews.com
http://www.merck.com
http://www.judicialwatch.org
http://www.naturalnews.com
http://science.naturalnews.com

Vaccination Debate

Much is being written about vaccinations these days. Accusations are flying left and right as for mandatory vaccinations or not. How does one know who is right or who is wrong in this particular topic. I decided to investigate the history of vaccinations. It is surprising what has been reported. Finally, there was a summit conference on vaccinations. Surprisingly, none of the vaccine proponents attended this first ever conference on vaccination safety summit.

The big reason for this discussion all of a sudden regarding mandatory vaccinations for everyone is a result of the measles outbreak in Disneyland in California. So let's discuss some of the reported events that triggered this outbreak. Right of the bat, without any evidence, a mythical, unvaccinated individual was blamed for the outbreak, but that is what is expected from the biased corporate news media presently. The number of cases grew rapidly with a large number of cases based in California but also having spread to other states. One must remember that Disneyland is a theme park that attracts millions of visitors into their facility every year from all over the world. No wonder that an infected individual, who is contagious, had infected so many people. It is interesting to note that the infected individuals were both unvaccinated and vaccinated individuals. This is significant. This signifies one thing—**the measles vaccine does not work**, and if it does, it is only for a short duration.

Brief history of vaccinations

In 1940, in Germany, compulsory mass vaccination was done against diphtheria. By 1945, diphtheria cases were up from 40,000 to 250,000. (Don't get stuck, Hannah Allen.)

In the USA in 1960, two virologists discovered that both polio vaccines were contaminated with the SV40 virus, which causes cancer in animals. SV40 virus is classified as polyoma—*poly* meaning "many," *oma* meaning "tumor" (*Med. Jnl. of Australia* 17/3/1973; p. 555).

In 1967, Ghana was declared measles-free by WHO. In 1972 Ghana had the highest measles outbreak with a corresponding highest mortality rate. (Dr. H. Albonico, MMR Vaccine Campaign in Switzerland, March 1990).

Between 1970 and 1990, the UK reported over 200,000 cases of whooping cough that occurred in vaccinated children (Community Disease Surveillance Centre, UK).

In 1970 in India, tuberculosis vaccine trial involved 260,000 people. It revealed that more cases of tuberculosis occurred in vaccinated than the unvaccinated (*The Lancet*).

In 1977, Dr. Jonas Salk, developer of the first polio vaccine, testified along with other scientists, that mass vaccination against polio was the most polio cases throughout the USA since 1961 (*Science* 4/4/77 "Abstracts").

In 1978, survey of thirty-nine states in the USA revealed that more than half of the children who came down with measles had been adequately vaccinated (the People's Doctor, Dr. R. Mendelsohn).

In 1979, Sweden ceased vaccination for whooping cough due to its ineffectiveness. In 1978, out of 5,140 cases, 84% had been vaccinated three times (*BMJ* 283:696-697, 1981).

In February 1981, an issue of the *Journal of the American Medical Association* found that 90% of obstetricians and 66% of pediatricians refused to take the rubella vaccine.

In 1982, a single vaccine shot for DPT cost 11 cents. In 1987, the cost rose to $11.40. The manufacturers of the vaccine had to place aside $8 per shot to cover legal fees and damages they were paying to parents of brain damaged children or children that died from the vaccination (*The Vine*, issue 7, January 1994, Nambour).

In Oman, between 1988 and 1989, a polio outbreak occurred among thousands of fully vaccinated children. The region with the highest attack rate had the highest vaccine coverage. The region with the lowest attack rate had the lowest vaccine coverage (*The Lancet*, 21/9/91).

A UK 1990 survey among 598 doctors revealed that over 50% of them refused to have the hepatitis B vaccine despite belonging to the high-risk group urged to be vaccinated (*British Med Lnl*, 27/1/1990).

In 1990, the *Journal of the American Medical Association* had an article on measles that stated, "Although more than 95% of school-age children in the US are vaccinated against measles, large measles outbreaks continue to occur in schools and most cases in this setting occur among previously vaccinated children" (*JAMA*, 21/11/90).

In USA from July 1990 to November 1993, the US Food and Drug Administration counted a total of 54,072 adverse reaction following vaccination. The FDA admitted that this number represented only 10% of the real total because doctors were refusing to report vaccine injuries. In other words, adverse reactions for this period exceeds half a million! (National Vaccine Information Centre, March 2, 1994).

In the *New England Journal of Medicine*, 1994 issue, a study found that over 80% of children under five years of age who had contracted whooping cough had been fully vaccinated.

On November 2, 2000, the Association of American Physicians and Surgeons (AAPS) announced that its members voted at their fifty-seventh annual meeting in St. Louis to pass a resolution calling for an end to mandatory childhood vaccines. The resolution passed without a single "no" vote (report by Michael Devit).

History of vaccines

In 1954, John F. Enders and Dr. Thomas C. Peebles collected blood samples from infected individuals with measles. They succeeded in isolating the measles virus from an infected teenager. The first measles virus vaccine was made by John Enders and colleagues, from a transformed strain named Edmonton-B and licensed in the US It consisted of a weakened measles virus. The vaccine was further changed in 1968 by a more weakened measles virus. This vaccine measles strain of Edmonton-Enders is the only measles vaccine used in the US. The measles vaccine was combined with mumps and rubella vaccine—the MMR vaccine. The measles cases were on a decline, and when measles vaccination was started, the elimination of measles elimination was accomplished in the year 2000.

A study of a ten-year period from 2003 on indicated no deaths due to the measles disease, yet deaths occurred during the same time frame due to measles vaccination. Centers for Disease Control and Prevention (CDC) keeps a weekly record of disease outbreaks, including death under the title CDC Morbidity and Mortality Weekly Reports (MMWR). These are all public records. Dr. Ann Schuchat, the director of CDC's National Center for Immunization and Respiratory Diseases, reported in an Associated Press article, "There have been no measles deaths reported in the US since 2003." Also, the US Government keeps a database under the heading "The Vaccine Adverse Reporting System" (VAERS). A search by Health Ranger, who publishes Natural News, for that same ten-year time period, revealed that deaths were reported under measles vaccination. According to their report, 108 deaths were recorded over that time frame, associated with four different measles vaccines used in the United States. Today, the measles vaccine comes in a combination with mumps and rubella (MMR). You can access these records at MedAlerts.com. Another place to search vaccine injuries or deaths is to look at US government settlements for MMR vaccine. You are probably unaware that manufacturers of vaccines have been given immunity from prosecution, but redress can be obtained by suing the US Government in a special "vaccine court."

http://www.naturalnews.com/048573_measles_deaths_MMR_vaccine_immunization_dangers.html#bczz3aJpqFG, http://www.medalerts.org

http://vaccineimpact.com

http://www.uscfc.uscourts.gov

http://www.historyofvaccines.org/content/blog/thomas-peebles-doctor-who . . .

http://www.washingtonpost.com/wp-dyn/content/article/2010/08/13/AR.

A study published in 1987 in *New England Journal of Medicine* reported on a measles outbreak in Corpus Christi, Texas, in the spring of 1985. Fourteen students contracted the measles, but what was interesting was all had been previously vaccinated. This contradicts the official statement that unvaccinated individuals are responsible for propagating measles outbreaks. This so-called herd immunity due to vaccinations propagated by health officials is nothing but a myth.

In 1988, the CDC published data on measles which documented 3,655 cases of measles in 1987. 1903 of these cases were in vaccinated individuals, or about 52%. Does this show the effectiveness of the MMR vaccine? The so-called antivaxxers (people who oppose vaccination) have a legitimate concern regarding effectiveness of vaccinations. This requires a further investigation into the world of vaccinations. Are they safe as purported by the pharmaceutical companies, FDA, CDC, and the medical profession?

The surprising amount of data obtained on individuals from the Disneyland outbreak indicates that only 14% of the sick individuals were unvaccinated. That means that the other 86% individuals were fully, up-to-date vaccinated. This clearly demonstrates that the MMR vaccination does not work as intended but does put the general population at risk. According to Ellison from the People's Chemist: "Your immune system is programmed to recognize and attack invaders that come through the biological 'front door'. That would be your nose, mouth, and eyes. It doesn't work properly when we shove infection into our body with a needle." Concerning herd immunity, Ellison adds that it's "nothing more than a silly catchphrase used to scare and bully parents into vaccinating their kids."

http://www.naturalnews.com/049351_measles_outbreak_MMR_vaccine_Disneyland.html#jxzz3aKF9ngU2

http://www.westonprice.org,

http://www.thepeopleschemist.com

In 1985, a measles outbreak in Corpus Christi occurred on vaccinated students who had up-to-date vaccination record. This was reported by in the *New England Journal of Medicine.* Researchers noted also that 99% of students at the school had been vaccinated. The students who were vaccinated should have been immune to the measles outbreak in their school. This verifies that the popular medical profession statement that vaccines are necessary is a total myth. The vaccines did not work as intended or are of a very short duration in effectiveness.

The dark side of vaccination history is that there are no safety studies done prior to releasing for general immunization. Finally, in early 2015, a Vaccination Safety Summit was organized. The only medical professional people attended this summit were doctors who did historical studies on vaccinations. No representatives from the pharmaceutical companies, FDA, or CDC attended this summit.

http://vaccineworldsummit.com

One of the more interesting fresh insights into vaccine debate was provided by Dr. Jane Orient, MD, director of the Association of American Physicians and Surgeons (AAPS): "Doctors are really intimidated" and when they do not fall in line with what the system says, it could spell the end of their medical career. "It's demeaning and it's insulting, but it's very, very powerful," she states. The system keeps an electronic tab on doctors, monitored by the higher-ups, to determine how the doctors are following the vaccination schedule for their patients.

http://vaccineworldsummit.com,

http://www.naturalnews.com/049009_Vaccine_World_Summit_Jane_Orient_mandatory_vaccination.html

Another bit of news from the world of vaccinations, two former virologists that worked for Merck, filed a lawsuit, accusing the pharmaceutical giant of marketing multivalent MMR vaccines under false pretenses. According to the lawsuit, these vaccines have been mislabeled, misbranded, adulterated, and falsely certified as having a 95% efficacy rate. The lawsuit meticulously details how Merck manipulated test results for two decades to falsely create the 95%–98% efficacy rate for the mumps component or their multivalent MMR vaccines. The reduced efficacy rate is directly responsible for mumps outbreak during the last decade according to the suit. The lawsuit claims that to this date (filed in 2010 and an amended complaint in 2012), Merck has consistently misrepresented the potency since 2000, by quoting the forty-year-old data from pre-MMR monovalent mumps vaccine, thereby misrepresenting the efficacy of four multivalent vaccines, namely MMR, MMRII, Europe's MMRVaxpro, and ProQuad, which is MMR plus chickenpox. The former employees also claimed, while employed by Merck, that they were threatened with jail, if they complained to FDA.

http://www.naturalnews.com/036298_Merck_scientists_False_Claims_Act_.html#jxzz3aVKavt5X

http://www.sanevax.org/wp-content/uploads/2012/06/Merck-mumps-suit.pdf

Two additional doctors that have dug into vaccine safety history are Drs. Suzanne Hunphries and Robert Rowan. According to Dr. Rowan, "Vaccines provide only a temporary immunity, while simultaneously and significantly increasing the risk of immune dysfunction, behavioral disorders and other major health problems. There is a graph of how these communicable diseases have fallen since the introduction of vaccines, and a corresponding, parallel, identical rise in chronic immune dysfunction, like asthma, arthritis, multiple sclerosis, and others. No one has ever done an all-cause morbidity and mortality study on the effectiveness and safety of vaccines." Dr. Rowan cites a study published in the *Journal of the American Medical Association* (*JAMA*) in 2010 that highlights a doubling in the rate of chronic health conditions among children between the years of 1994 and 2006. It went from 12.8% to 26.6%. This directly corresponds to a substantial increase in the number of vaccines added to the government's vaccine program.

http://www.naturalnews.com/049007_Robert_Rowan_Vaccine_World_Summit_immunizations.htm;#jxzz3Uh2Ao78C

Following are some interesting graphs regarding some of our vaccination history:

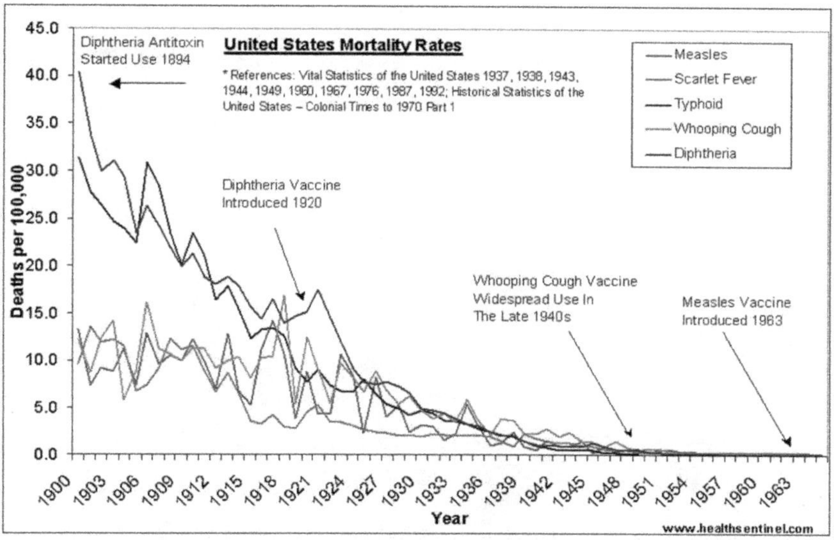
United States Mortality Rates
* References: Vital Statistics of the United States 1937, 1938, 1943, 1944, 1949, 1960, 1967, 1976, 1987, 1992; Historical Statistics of the United States – Colonial Times to 1970 Part 1
Diphtheria Antitoxin Started Use 1894
Diphtheria Vaccine Introduced 1920
Whooping Cough Vaccine Widespread Use In The Late 1940s
Measles Vaccine Introduced 1963
Measles
Scarlet Fever
Typhoid
Whooping Cough
Diphtheria
Deaths per 100,000
Year
www.healthsentinel.com

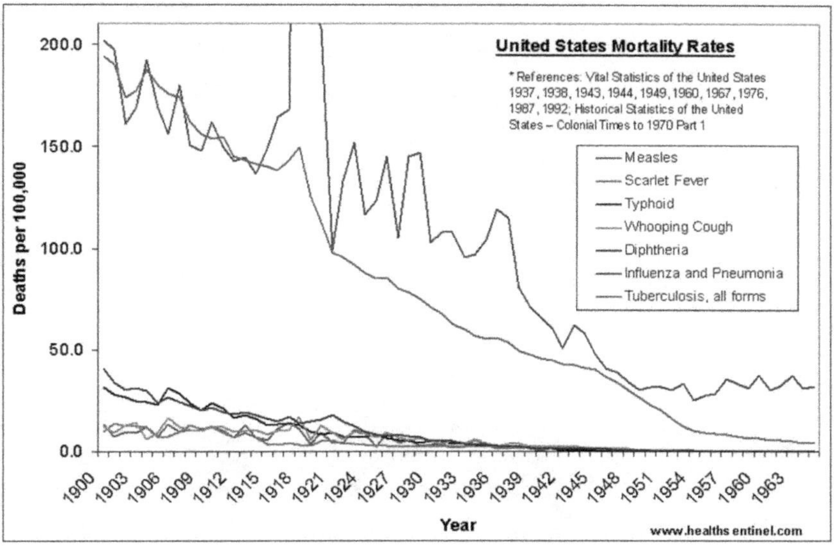
United States Mortality Rates
* References: Vital Statistics of the United States 1937, 1938, 1943, 1944, 1949, 1960, 1967, 1976, 1987, 1992; Historical Statistics of the United States – Colonial Times to 1970 Part 1
Measles
Scarlet Fever
Typhoid
Whooping Cough
Diphtheria
Influenza and Pneumonia
Tuberculosis, all forms
Deaths per 100,000
Year
www.healthsentinel.com

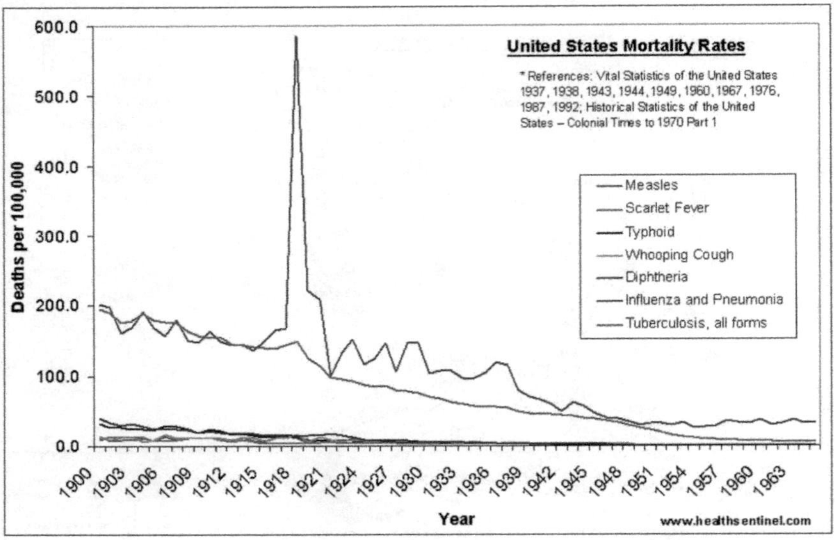
United States Mortality Rates
* References: Vital Statistics of the United States 1937, 1938, 1943, 1944, 1949, 1960, 1967, 1976, 1987, 1992; Historical Statistics of the United States – Colonial Times to 1970 Part 1
Measles
Scarlet Fever
Typhoid
Whooping Cough
Diphtheria
Influenza and Pneumonia
Tuberculosis, all forms
Deaths per 100,000
Year
www.healthsentinel.com

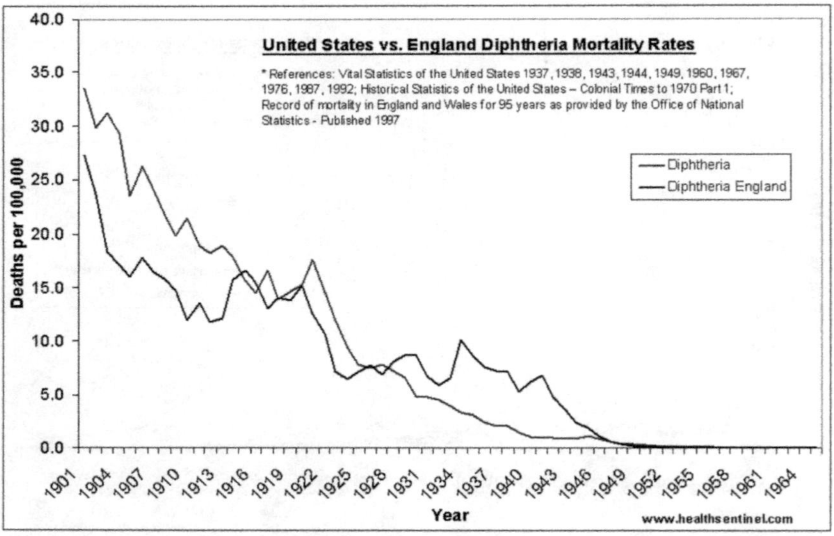
United States vs. England Diphtheria Mortality Rates
* References: Vital Statistics of the United States 1937, 1938, 1943, 1944, 1949, 1960, 1967, 1976, 1987, 1992; Historical Statistics of the United States – Colonial Times to 1970 Part 1; Record of mortality in England and Wales for 95 years as provided by the Office of National Statistics - Published 1997
Diphtheria
Diphtheria England
Deaths per 100,000
Year
www.healthsentinel.com

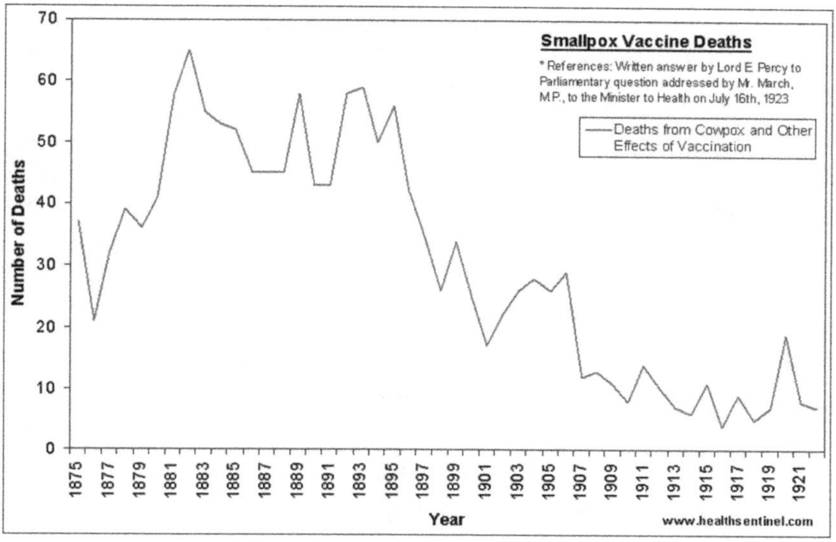

Another doctor who decided to delve into vaccine history and safety is Dr. Suzanne Humphries. She decided to research the histories of the smallpox and polio vaccines. According to her, "in my research, I was startled that what I found was completely counter to what I have been told and taught my entire life. I now don't believe that smallpox vaccines eradicated smallpox. I now don't believe that polio vaccines eradicated polio . . . it is easy to throw up smoke screens here and there and make whatever argument one might want to, because **people are ignorant** and because the story is so complicated."

The smallpox vaccine invented by Edward Jenner was basically scraping the pus off the belly of a cow. This was due to rumor mill at the time that the milkmaids infected with the cowpox would be immune to the smallpox. The vaccine consisted of puncturing the skin several times containing a mixture of the virus and glycerin. What the history of this vaccine indicated is that there were many vaccinated people who developed serious smallpox disease. Once the vaccination was eliminated, the disease incidence went down as it was proven out in Leicester town in England.

The other prime argument used for vaccine propagandists is the history of polio vaccine. According to Dr. Humphries, "the historical perspective fails to support the vaccinations paradigm." The diagnosis used to determine polio was based on two physical examinations within twenty-four hours. If the examinations indicated paralysis in one or more muscle, it was deemed to be from polio. Now it has been determined that a number of viruses can cause paralysis. The polio vaccine injection consisted of live poliovirus, thought to be inactivated with addition of formaldehyde. Since the formaldehyde did not inactivate the polio virus, more people developed paralysis from the vaccine than from normal wild, natural poliovirus. According to Dr. Humphries, "something had to be done to make it appear as though the vaccine was working . . . A change was made in the diagnostic criteria for polio . . . With polio, the original criteria was two examinations within 24 hours. This was changed to two examinations within 60 days . . . because within 60 days, most people recovered from their bout with poliomyelitis." Therefore, if the observation does not fit one's expectations, a change is made or rig the system so that they do. "all those people who were formerly called polio were no longer categorized as polio because they recovered from their paralysis within that time." Dr. Humphries published a book titled *Dissolving Illusions: Disease, Vaccines, and the Forgotten History.*

She also mentions that "people have been scared into believing vaccines are the answer to prevent disease, but when you look at the historical evidence, the arguments used simply fall apart . . . We have a highly profitable, lucrative, religion that involves the government, industry, and academia. That religion is vaccination."

Medical schools do not teach the history of vaccinations. They are brainwashed as to the effectiveness, and in almost all instances they are not taught the adverse health effect of vaccination. The adverse health effects are well documented, but there are only a handful of real doctors who delve into the history of vaccination. The medical-pharmaceutical mafia threatens them with suspension of their medical practice license if they go against the vaccination protocol.

http://www.articles.mercola.com/ . . . /aechive/2015/01/18/history-vaccinations.aspx,

Another interesting development was that a number of researchers suspected that the polio vaccine was contaminated with the SV40 virus, which caused tumors in humans. According to a report by American News, evidence of SV40 infections were showing up in children born after 1982. In 2002, the British Journal Lancelot published evidence of polio vaccine contamination that was responsible for a large number of instances of non-Hodgkin's lymphoma disease.

http://www.americanlivewire.com/AmericanNews

There is a report of an interview with Dr. Maurice Hilleman, a leading vaccine pioneer. While employed by Merck, he is responsible for more than three dozen vaccines. He was responsible for the vaccine program at Merck. In the interview, Dr. Hilleman openly admits that vaccines given to citizens here in America were contaminated with leukemia and cancer viruses. Dr. Hilleman was also responsible for importing the African green monkeys into the United States, not knowing that these monkeys contained the AIDS virus, and that is how AIDS was introduced in the US. Since the vaccines contained leukemia and AIDS viruses, they became pandemic from these wild viruses brought into this country, but it was "good science," according to Dr. Hilleman.

http://www.naturalnews.com/033584_Dr_Maurice_Hilleman_SV40.html

http://www.infowars.com/merck-scientist-dr-maurice-hilleman-admitted

The measles debate has quickly turned into a political tug of war, but what is not reported by the mainstream media is that the vaccinated individuals with the measles vaccine are more dangerous than those who are unvaccinated. There are numerous published studies that show that individuals who have been vaccinated with the MMR vaccine shed the disease for weeks or even months. These individuals could be infecting other people and are especially dangerous around individuals with compromised immune system. CDC realized this phenomenon when they tested urine samples from fifteen-month-old children and also teens and found nearly all of them had detectable measles virus in their bodies. The vaccine also contains human DNA from fetal cells linked to autism. The study's lead researcher, Helen Ratajczak, former scientist at a pharmaceutical company, stated, "Documented causes of autism include genetic mutations and/or deletions, viral infections, and encephalitis following vaccination. Therefore, autism is the result of genetic defects and/or inflammation of the brain."

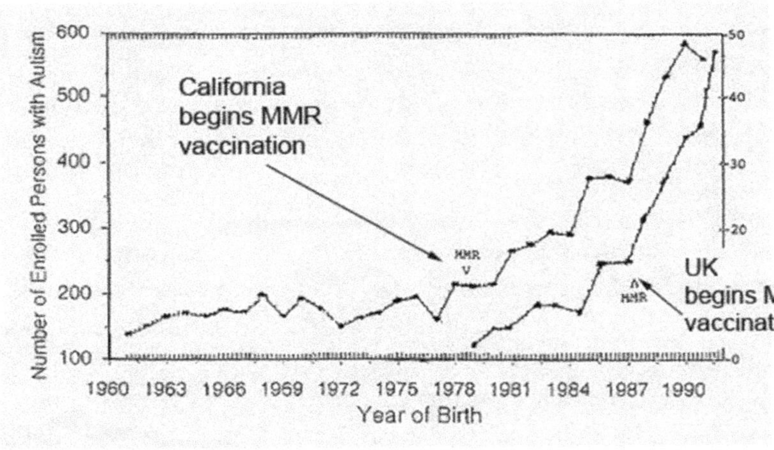

http://www.healthimpactnews.com
http://www.thenewamerican.com
http://www.historyofvaccines.com
http://www.naturalnews.com/048791_MMR_vacines_aborted_babies_human_DNA.html#jxzz3XRX68bQW

One of the adverse health effects of the measles vaccine is a potential progressive neurological disorder in children and young adults that affect the central nervous system. It is subacute sclerosing panencephalitis (SSPE). It is a slow, persistent viral infection due to a defective measles virus. There is a long period between measles onset and the development of SSPE. The initial symptoms of SSPE are mild, mental deterioration, changes in behavior, followed by disturbances in motor function, involuntary movements of the head, trunk, or limbs. Seizures may also occur. Progressive deterioration leads to comatose and finally death. Prevention in the form of measles vaccination is the only cure for SSPE.

http://www.ninds.nih.gov/Sisorders

The real story about vaccines begins when Congress in 1986 established the National Childhood Vaccine Injury Act. Under this law, Congress took the power of a lawsuit against pharmaceutical companies that manufacture vaccines, away from families. In essence, it established a vaccine court, which is nothing more than a federal claims court that deals specifically with vaccine cases, where families can go for injury compensation due to a vaccine injury. The official name is Vaccine Injury Compensation Program (VICP). This program was necessary since every child had to be vaccinated in order to attend preschool, day care, or private school. When VICP was created, it provided immunity to vaccine manufacturers or health care practitioners. The petitions for redress had to be filed against HHS (Health and Human Services). The other provision in the 1986 law permits the vaccine makers the right to not disclose known risks to parents or guardians of those being vaccinated. (What a perfect vehicle for globalists that want to decrease the population.) Families of injured individuals would then be compensated through an administrative process based on a table of presumptive injuries.

Since the advent of this vaccine court, it has paid out three billion dollars to settle claims to families of vaccine injured children. Neurological damage due to vaccination is not a rare occurrence. You don't read about this in the mainstream prosti-media. Yet they still propagandize vaccinations. A recent story came from United Kingdom, where GlaxoSmithKline (GSK) paid sixty-three million to settle a lawsuit filed against them due to their swine flu vaccine adverse health effects. Peter Todd, attorney representing the victims, stated, "There has never been a case like this before. The victims of this vaccine have an incurable and lifelong condition and will require extensive medication." The Global Research Project reports, the GlaxoSmithKlein rap sheet states, *"In recent years, GlaxoSmithKlein has become known as the company that pays massive amounts to resolve wide-range charges brought by US regulators and prosecutors. These included a $750 million payment relating to the sale of adulterated products from a facility in Puerto Rico and a record $3 billion in connection with charges relating to illegal marketing, suppression of adverse* *safety research results and overcharging government customers. The company also set a record for the largest tax avoidance settlement with the US Internal Revenue Service."*

Another recent story came from Mexico. Two children are dead and twenty-nine others are in critical condition after they had received a series of vaccines for tuberculosis, rotavirus, and hepatitis B. Around 75% of the 52 children vaccinated suffered severe reactions and two died. According to Newswatch Report, the vaccines in question were administered as a single triple vaccination, manufactured by Merck and Co. Merck's Rotateq vaccine for rotavirus contains DNA from two porcine circoviruses PCV1 and PCV2, which FDA has warned against since 2010. PCV2 is a lethal virus that causes immune suppression and serious wasting disease in baby pigs that damages lungs, kidneys, reproductive system, and brain and ultimately causes death.

http://naturalsociety.com/brain-damaged-victims-of-swine-flu-vaccine-win-63-million-lawsuit/#jxzz3Xm3MjJIO

http://www.naturalnews.com/049719_Merck_vaccines_Rotateq.html#jxzz3aJPRibJb,

http://latino.foxnews.com,

http://newswatchreport.com

How about the woman who was planning a trip with her husband to Africa? The required shots included tetanus, typhoid, and hepatitis A and B vaccinations. Two days later, she was completely debilitated, a victim of permanent brain damage from encephalopathy. This is a person who once was a mayor of a town, reduced to almost vegetative state due to vaccination. She received 7.4 million dollars in damages since she requires round-the-clock care. Yet CDC and mainstream media continue to push for vaccination, claiming that they are safe.

http://naturalnews365.com/vaccine-dangers-brain-damage-1352.html#sthash.z9WIPOLv.dpuf

So what do our illustrious representatives that represent us in Washington, DC, do. They held a congressional hearing. The National Center for Immunization and Respiratory Disease Director, Dr. Ann Schuhat, tells Congress that "there is absolutely no scientific evidence that vaccines cause autism, mental disorders, allergies or autoimmune disease." How can someone in authority lie like that and our elected officials believe these lies? **And to further these blatant lies, Massachusetts senator Elizabeth Warren recently proposed eliminating all criminal penalties for drug companies that injure or kill people with their deadly products. And this person wants to be president? There are hundreds of studies linking vaccines to autism, neurological problems, autoimmune suppression, allergies, and numerous mental problems as well as death.** The first court award, admitting that vaccines cause autism, was 2010. Family of Hannah Poling received $1.5 million for lost earnings, pain, and suffering and life care. The court agreed that Hannah developed autism after receiving nine vaccines, including MMR vaccine, all at once.

http://www.naturalnews.con/048096_CDC_vaccine_injury_perjury.html#lxzz3UHPN9WXu.

Another case involving vaccine injury is reported that a six-month-old baby boy, after receiving thirteen vaccinations at once. These were two triple doses of DTaP, hepatitis B, a polio shot, three oral rotavirus doses, and a pneumococcal pneumonia vaccine, according to the mother, compressed into three shots and one oral dose. The boy was perfectly healthy prior to the vaccination. Five days later, he was dead.

A study regarding infant mortality in 2010 determined that US has the highest rate among the twenty-eight wealthy industrial nations—6.1 out of 1,000. The US has the highest number of vaccines given to infants. There is an organization called Angel Babies, a support group for parents who have children that died following vaccinations. Medical professionals continuously write off infant death after vaccination as "sudden infant death syndrome." Actual evidence from scientific and clinical literature produced by CDC and FDA contradicts their statements that vaccines are safe. A new study was published in the *Journal of Pediatrics* titled "Adverse Events Following *Haemophilus influenzae* Type b Vaccines Adverse Event Reporting System," 1990–2013. In this study, CDC and FDA researchers identify 749 deaths linked to the administration of the Hib vaccine, where

51% were sudden infant death (SIDS) linked to that vaccine. SIDS (sudden infant death syndrome) is the third largest cause of death in infants. A top CDC scientist has finally exposed the neglect and suppression of an autism-MMR link in African American boys. He was compelled to cover-up this data but decided to come forward with the information. The CDC has continually denied of any evidence supporting sudden infant death, but do state the following: "From 2 to 4 months old, babies begin their primary course of vaccinations. This is also peak age for sudden infant death syndrome (SIDS)." CDC usually writes this as coincidence. In a new study the CDC and FDA researchers acknowledge scarcity of post licensure safety data on Hib vaccine. The study uncovered the following highly concerning results reported to VAERS (Vaccine Adverse Event Reporting System). This agency received 29,747 reports after Hib vaccination. 5,179 were serious, including 896 deaths. SIDS was the stated cause of death in 384 cases with autopsy/death certification records. One has to be aware that VAERS is a passive surveillance system, which in all probability suffers from profound underreporting. The researcher's conclusion: "Review of VAERS reports did not identify any new or unexpected safety concerns for Hib vaccine." What a troubling disregard for human/infant life. These federal agencies misrepresent the safety over human wellbeing. How can such ignorant so-called researchers make statements like that, since the data proves them wrong. Maybe CDC should stand for Centers of Death Control and FDA for Federal Death Agency. A very interesting bit of history regarding vaccinations comes from Australia, where in the 1960s and 1970s Aboriginal babies began mysteriously die at an almost 50% rate. The riddle was solved when an Australian physician, Dr. Archie Kalokerinos, realized that the deaths occurred after the infants were vaccinated. These babies were severely undernourished. Their systems could not handle the stress placed on their immune system. In Japan, vaccination date was raised from three months to two years, and the corresponding SIDS rate dropped from 12.4 to 5.0. Cherry et al. made this observation: "It is clear that delaying the initial vaccination until a child is 24 months, regardless of the type of vaccine, reduces most of the temporary associated severe adverse reactions."

There are numerous studies presented by large contingent of doctors at the annual meetings of American Academy of Pediatrics showing that two-thirds of babies that died from SIDS had been vaccinated against DPT prior to death. The death occurred from twelve hours after vaccination and up to three weeks later. According to a study done by William C. Torch, MD, vaccinated babies died most often at two to four month, coinciding when initial doses of DPT were given to the infants. He concluded, "DPT may be a generally unrecognized major cause of sudden infant and early childhood death, and that the risks of immunization may outweigh its potential benefits. A need for re-evaluation and possible modification of current vaccination procedures is indicated by this study."

http://greenmedinfo.com/blog/cdcs-own-data-vaccine-infant-death-link-56k

http:/www.naturalnews.com/042727_infants_sudden_death_vaccines.html#jxzz3UpBhY6C2

One of the more thorough discussions regarding adverse health effects on the developing fetal and early infant brain is in this link:

http://www.article.mercola.com/ . . . danger-of-excessive-vaccination-during-brain-development.aspx

What is now being proposed is mandatory vaccinations of all individuals in the country. This tendency of mass vaccination is to herd people, like the fluoridation of water with a poison like fluoride. In each instance there are no studies proving any health benefits, but on the contrary, numerous studies listing adverse health effects and death due to vaccination. All one has to do is access the history of fluoride and specifically vaccinations to object to this mandatory medicine on the entire population. For a long time, the primary targets for total vaccinations were the newborn children. The amount of vaccines targeting the infants rose increasingly to over thirty by the time they are two years old. The number of adverse health effects on our children increased with increased vaccine administration.

Now it has come to our attention that the government has an agenda titled National Adult Immunization Plan (NAIP). This is supposed to be a five-year plan to push adults into getting more vaccines than ever before. HHS instead addressing concerns about neurotoxins, fetal cells, and other toxic dangerous chemicals in the vaccine, they want to create electronic health records to monitor populations from the inside. This is contrary to our right of privacy. This move toward mandatory vaccination is a move toward medical rights suppression or could this also be population control? With this NAIP and the government propaganda and Big Brother medical surveillance of adult vaccinations are definitive steps in that direction.

http://www.hhs.gov/nvpo/national_adult_immunization_plan_draft.pdf

http://jonrappoport.wordpress.com/2015/02/25/adu;t-i,, immunization-push-medical-dictatorship

http://www.naturalhealth365.com/adu;t=vaccination-schedule-1347.html#sthash.POG1ttkg.dpuf

The problem that will arise from this mass vaccination hysteria is that it would lead to a totalitarian state, alas what George Orwell wrote in his book 1984, where big brother is watching you. Medical critics are very clear on this: **no one should be forced to undergo a medical treatment without informed consent or agreement to the treatment.** The vaccine war is nothing more than a culture war on freedoms, values, and beliefs that have long defined who we are as a nation. Adult immunization plan targets pregnant women, employers and their employees, faith groups, etc. Most health care employers already subjected their employees to mandatory vaccines as a condition of their employment. With such a plan you should ask yourself the following questions: Can I get a driver's license without vaccination? Can I get health care coverage without vaccination? Can I get Social Security benefits without vaccinations? To name a few. All these freedoms can be wiped out when that becomes a requirement by those agencies. If that law is passed by our ignorant legislators, the following loss of liberty follows. The problem stems from that some of our legislators are in the pockets of the biotech and pharmaceutical industries and as such listen to their demands.

The population at large is being bombarded with the following vaccination myths:

1) **Most doctors say that vaccines do not cause injury or death, so it must be true.** The problem with this statement is that vaccinations are taught in medical schools as safe; therefore, you have a large body of health care providers who are misinformed and ignorant as to the adverse health effects of vaccines. Knowing this, Congress in 1986 and the Supreme Court in 2011 declared vaccines to be **unavoidably unsafe** but shielded the drug companies from all product liability. The problem with such a law is that it precludes anyone from suing the vaccine manufacturers in a civil lawsuit. They established a vaccine court instead.
2) **People who question vaccine safety are ignorant and do not understand the science.** The opposite is true. People who question the vaccine safety do their homework by investigating all the scientific studies that do exist to make an intelligent decision regarding vaccination. These people question the validity of the increased amount of vaccinations that the children are subjected to in their early age. Two decades of reports published by the Institute of Medicine found large knowledge gaps on vaccines.
3) **People who do not vaccinate are selfish and don't care about public health.** Many people do not vaccinate because the risks of adverse health effects like allergies, brain damage, autism, autoimmune disease, etc., have resulted according to good independent scientific studies. No safety data is provided to vaccinated individuals except a word from the misinformed health provider. Some people choose a different path to boost their immunity through better nutrition like organic foods, exercise, and avoiding environmental toxins. They also do not rely on heavy use of pharmaceutical products. Also they prefer to do informed choices regarding their health. Some mothers opt out of vaccinating their subsequent offspring after their first infant suffered adverse health effect from vaccines.

4) **Science trumps religious beliefs, so religious exemptions to vaccination should be eliminated.** The attack on Americans with religious beliefs is paramount in the present political arena. This political stance is being driven by individuals who do not believe in God. This type of thinking goes beyond religious belief, with goal to demoralize the youth of the country and making them subservient to the government utilitarian policies.

5) **It is ethical for government to sacrifice the few for the many.** This philosophy is utilitarianism. The utilitarian rationale was created as a guide for legislators making public policy. This type of thinking was given a green light with the decision in 1923 by US Supreme Court Justice Oliver Wendell Holmes when he ruled that the state of Virginia had a right to sterilize a young woman because doctors judged her to be mentally retarded just like her daughter and mother. This utilitarian philosophy was rampart under the Nazi regime where horrific medical experiments were performed on captive people. The Nuremberg trial declared utilitarianism to be pseudoethical. They issued the consent principle, which has guided research on humans and the ethical practice in medicine. This apparently is completely ignored by our so-called health guardian federal agencies and the medical and pharmaceutical companies presently, since we have become wholesale guinea pigs for the medical profession. Forced vaccinations sacrifice the genetically vulnerable individuals since they possess unknown genetic risk factors. The other problem with utilitarianism is that the medical industry, agricultural industry and pharmaceutical industries with government consent refuse to fund good independent science to better understand biological mechanism. Eliminating nonvaccination objection exemptions will guarantee the elimination of health outcome between vaccinated and unvaccinated individuals. To learn more you can access the NVIC Memorial for

Vaccine Victims where they contain a wealth of information regarding vaccine injuries. Be informed and access the links provided.
http://vaccines.mercola.com/sites/vaccines/download-nvic-posters.aspx

Doctors and scientists condemn vaccination

"There is a great deal of evidence to prove that immunization of children does more harm than good" (Dr. J Anthony Morris, former chief vaccine control officer, US Food and Drug Administration).

"The greatest threat of childhood disease lies in the dangerous and ineffectual efforts made to prevent them through mass immunization" (Dr. R. Mendelsohn, author and professor of pediatrics, *How to Raise a Healthy Child in Spite of Your Doctor*).

"In our opinion, there is now sufficient evidence of immune malfunction following current vaccination programs to anticipate growing public demands for research investigation into alternative methods of prevention of infectious disease" (Drs. H. Buttram and J. Hoffman *Vaccinations and Immune Malfunctions*)

"All vaccination has the effect of directing the three values of the blood into or toward the zone characteristics of cancer and leukemia . . . Vaccines DO predispose to cancer and leukemia" (Professor L. C. Vincent, founder of bioelectronics).

"Every vaccine carries certain hazards and can produce inward reactions in some people . . . in general, there are more vaccine complications than is generally appreciated" (Professor George Dick, London University).

"Official data have shown that the large-scale vaccinations undertaken in the US have failed to obtain any significant improvement of the diseases against which they were supposed to provide protection" (Dr. A. Sabin, developer of the oral polio vaccine, lecture to Italian doctors in Piacenza, Italy, December 7, 1985).

"In addition to the many obvious cases of mortality from these practices, there are also long-term hazards which are almost impossible to estimate accurately . . . the inherent danger of all vaccine procedures should be a deterrent to their unnecessary or unjustifiable use" (Sir Graham Wilson, *The Hazards of Immunization*).

"Laying aside the very real possibility that the various vaccines are contaminated with animal viruses and may cause serious illness later in life (multiple sclerosis, cancer, leukemia, etc.) we must consider whether the vaccines really work for their intended purpose" Dr. W. C. Douglas, *Cutting Edge*, May 1990).

"The only wholly safe vaccine is a vaccine that is never used" (Dr. James A. Shannon, National Institute of Health, USA)

With reference to smallpox

"Vaccination is a monstrosity, a misbegotten offspring of error and ignorance, it should have no place in either hygiene or medicine . . . Believe not in vaccination, it is a worldwide delusion, an unscientific practice, a fatal superstition with consequences measured today by tears and sorrow without end" (Professor Chas Rauta, University of Perguia, Italy, *New York Medical Journal*, July 1899).

"Vaccination does not protect, it actually renders its subjects more susceptible by depressing vital power and diminishing natural resistance, and millions of people have died of smallpox which they contracted after being vaccinated" (Dr. J. W. Hodge, *The Vaccination Superstition*).

"It is nonsense to think that you can inject pus—and it is usually from the pustule end of the dead smallpox victim . . . it is unthinkable that you can inject that into a little child and in any way improve its health. What is true of vaccination is exactly as true of all forms of serum immunization, if we could by any means build up a natural resistance to disease through these artificial means, I would applaud it to the echo, but we can't do it" (Dr. William Howard Hay, lecture to Medical Freedom Society, June 25, 1937).

"Immunization against smallpox is more hazardous than the disease itself" (Professor Ari Zuckerman, World Health Organization).

With reference to whooping cough

"There is no doubt in my mind that in the UK alone some hundreds, if not thousands of well infants have suffered irreparable brain damage needlessly and that their lives and those of their parents have been wrecked in consequence" (Professor Gordon Stewart, University of Glasgow, *Here's Health*, March 1980).

"My suspicion, which is shared by others in my profession, is that the nearly 10,000 SIDS deaths that occur in the US each year are related to one or more of the vaccines that are routinely given to children. The pertussis (whooping cough) vaccine is the most likely villain, but it could also be one or more of the others" (Dr. R. Mendelsohn, author and professor of Pediatrics, *How to Raise a Healthy Child in Spite of Your Doctor*).

"The worst vaccine of all is the whooping cough vaccine . . . it is responsible for a lot of deaths and for a lot of infants suffering irreversible brain damage" (Dr. Archie Kalokerinos, author and vaccine researcher, Natural Health Convention, Stanwell Tops, NSW, Australia 1987).

With reference to polio

"Many here voice a silent view that the Salk and Sabin polio vaccine, being made of monkey kidney tissue has been directly responsible for the major increase in leukemia in this country" (Dr. F. Klenner, polio researcher, USA).

"No batch of vaccine can be proved to be safe before it is given to children" (Surgeon General Leonard Scheele, AMA Convention 1955, USA).

"Live virus vaccines against influenza and paralytic polio, for example, may in each instance cause the disease it is intended to prevent" (Dr. Jonas Salk, developer of first polio vaccine, *Science*, 4/4/77 Abstracts).

A very good source on vaccine articles that were written by some news media is by accessing the http://PEERS.WantToKnow site. Following are only two pages of articles regarding vaccinations.

How independent are vaccine defenders?
2008-07-25, CBS News
http://www.cbsnews.com/stories/2008/07/25/cbsnews_investigates/main4296175.shtml

> They're some of the most trusted voices in the defense of vaccine safety: the American Academy of Pediatrics, Every Child By Two, and pediatrician Dr. Paul Offit. But CBS News has found these three have something more in common—strong financial ties to the industry whose products they promote and defend. The vaccine industry gives millions to the Academy of Pediatrics for conferences, grants, medical education classes and even helped build their headquarters. The totals are kept secret, but public documents reveal bits and pieces. A $342,000 payment from Wyeth, maker of the pneumococcal vaccine—which makes $2 billion a year in sales. A $433,000 contribution from Merck, the same year the academy endorsed Merck's HPV vaccine—which made $1.5 billion a year in sales. Every Child By Two, a group that promotes early immunization for all children, admits the group takes money from the vaccine industry, too—but wouldn't tell us how much. **Then there's Paul Offit, perhaps the most widely-quoted defender of vaccine safety. He's gone so far as to say babies can tolerate "10,000 vaccines at once." In fact, he's a vaccine industry insider. Offit holds in a $1.5 million dollar research chair at Children's Hospital, funded by Merck. He holds the patent on an antidiarrhea vaccine he developed with Merck. And future royalties for the vaccine were just sold for $182 million cash**

A coverup for a cause of autism?
2005-06-22, MSNBC
http://www.msnbc.msn.com/id/8243264/ns/msnbc-morning_joe/t/coverup-cause-autism

JOE SCARBOROUGH, Host: Six out of every 1,000 kids get it, and nobody knows exactly why. But my next guest says . . . part of the blame . . . needs to fall on government. And it has to do with a drug called thimerosal. Robert F. Kennedy Jr. is a senior attorney for the Natural Resources Defense [Council]. Let's talk tonight about thimerosal. There are a lot of people out there . . . very concerned about the impact of this drug, which is found in vaccines, and how it causes autism. Talk about that. **ROBERT F. KENNEDY JR.:** That's right. Thimerosal is a preservative that was put in vaccines back in the 1930s. Almost immediately after it was put in, autism cases began to appear. Autism had never been known before. It was unknown to science. Then the vaccines were increased in 1989 by the CDC and by a couple of other government agencies. What happened was the vaccine schedule was increased. We went up from receiving about 10 vaccines in our generation to these kids receive 24 vaccines. And they all had this thimerosal in them, this mercury. And **nobody bothered to do an analysis of what the cumulative impact of all that mercury was doing to kids. As it turns out, we are injecting our children with 400 times the amount of mercury that FDA or EPA considers safe. A child on his first day that he is born is injected with a hepatitis B shot. Under EPA guidelines, he would have to be 275 pounds to safely absorb that shot.**

What happened was that, in 1988, one in every 2,500 American children had autism. Today, one in every 166 children have autism.

Vaccine virus "cancer link"
2002-03-08, BBC News
http://news.bbc.co.uk/2/hi/health/1860042.stm

A monkey virus found in early versions of a vaccine against polio may be linked to a common type of cancer, suggest scientists. Batches of polio vaccine tainted with "simian virus 40" (SV40) were given between 1955 and 1963. This was because monkey kidney cells were used in the [vaccine's] production process. It is [now] conceded that SV40 was present in the early vaccine—and the latest research, published in the *Lancet* journal, has linked it to non-Hodgkin's lymphoma. This is a cancer of the lymphatic system, which has a role in the body's fight against infection, and affects mainly the over 40s. The researchers looked at hundreds of tumours taken from various cancer patients, and compared them with 68 samples taken from non-Hodgkin's patients. They found genetic "footprints" of the virus in 43% of the non-Hodgkin's tumour cells.

Influenza: marketing vaccine by marketing disease
2013-05-16, *British Medical Journal*
http://www.bmj.com/content/346/bmj.f3037

The CDC pledges "To base all public health decisions on the highest quality scientific data." In the case of influenza vaccinations and their marketing, this is not so. **Promotion of influenza vaccines is one of the most visible and aggres-**

sive public health policies today. Although proponents employ the rhetoric of science, the studies underlying the policy are often of low quality, and do not substantiate officials' claims. The vaccine might be less beneficial and less safe than has been claimed, and the threat of influenza appears overstated. Twenty years ago, in 1990, 32 million doses of influenza vaccine were available in the United States. Today [the number is] around 135 million doses. This enormous growth has not been fueled by popular demand but instead by a public health campaign. Drug companies have long known that to sell some products, you would have to first sell people on the disease. In the 1950s and 1960s, Merck launched an extensive campaign to lower the diagnostic threshold for hypertension, and in doing so enlarging the market for its diuretic drug, Diuril. Could influenza . . . be yet one more case of disease mongering? Marketing influenza vaccines . . . involves marketing influenza as a threat of great proportions. The CDC's website explains that "Flu seasons . . . can be severe," citing a death toll of "3000 to a high of about 49000 people." However, a far less volatile and more reassuring picture of influenza seems likely if one considers that recorded deaths from influenza declined sharply over the middle of the 20th century . . . all before the great expansion of vaccination campaigns in the 2000s. Yet across the country, mandatory influenza vaccination policies have cropped up . . . precisely because not everyone wants the vaccination, and compulsion appears the only way to achieve high vaccination rates.

Scandal exposed in major study of autism and mercury
2011-10-25, *Sacramento Bee* (leading newspaper in California's capital city)
http://www.sacbee.com/2011/10/25/4005040/scandal-exposed-in-major-study.html

> The Coalition for Mercury-free Drugs (CoMeD) exposes communications between Centers for Disease Control (CDC) personnel and vaccine researchers revealing **US officials apparently colluded in covering-up the decline in Denmark's autism rates following the removal of mercury from vaccines. Documents obtained via the Freedom of Information Act (FOIA) show that CDC officials were aware of Danish data indicating a connection between removing Thimerosal (49.55% mercury) and a decline in autism rates.** Despite this knowledge, these officials allowed a 2003 article to be published in Pediatrics that excluded this information, misrepresented the decline as an increase, and led to the mistaken conclusion that Thimerosal in vaccines does not cause autism. In Denmark, Thimerosal, a controversial mercury compound used as a preservative in certain vaccines, was removed from all Danish vaccines in 1992. The well-publicized Danish study published in Pediatrics 2003 claimed that autism rates actually increased after Thimerosal was phased out. This study subsequently became a cornerstone for the notion that mercury does not cause autism. However, one of the FOIA documents obtained from CDC clearly indicates that this study omitted large amounts of data showing autism rates actually dropping after mercury was removed from Danish vaccines.

Measles among vaccinated Quebec kids questioned
2011-10-20, Canadian Broadcasting Corporation (Canada's NPR)
http://www.cbc.ca/news/health/story/2011/10/20/measles-quebec-vaccine-schedul . . .

> Measles cases have surged in parts of Canada and the United States this year. A still smoldering outbreak of measles in Quebec is the largest in the Americas in over a decade. An investigation into an outbreak in a high school in a town that was heavily hit by the virus found that about half of the cases were in teens who had received the recommended two doses of vaccine in childhood — in other words, teens whom authorities would have expected to have been protected from the measles virus. **It's generally assumed that the measles vaccine . . . should protect against measles infection about 99 per cent of the time. So the discovery that 52 of the 98 teens who caught measles were fully vaccinated came as a shock to the researchers who conducted the investigation.**

Vaccines and autism: a new scientific review
2011-03-31, CBS News
http://www.cbsnews.com/8301-31727_162-20049118-10391695.html

> For all those who've declared the autism-vaccine debate over – a new scientific review begs to differ. It considers a host of peer-reviewed, published theories that show possible connections between vaccines and autism. The article in the *Journal of Immunotoxicology* is entitled "Theoretical aspects of autism: Causes—A

review." The author is Helen Ratajczak, surprisingly herself a former senior scientist at a pharmaceutical firm. Ratajczak did what nobody else apparently has bothered to do: she reviewed the body of published science since autism was first described in 1943. Not just one theory suggested by research such as the role of MMR shots, or the mercury preservative thimerosal; but all of them. Ratajczak's article states, in part, that "Documented causes of autism include genetic mutations and/or deletions, viral infections, and encephalitis following vaccination. Therefore, **autism is the result of genetic defects and/or inflammation of the brain." The article goes on to discuss many potential vaccine-related culprits, including the increasing number of vaccines given in a short period of time. Ratajczak also looks at a factor that hasn't been widely discussed: human DNA contained in vaccines.** Ratajczak reports that about the same time vaccine makers took most thimerosal out of most vaccines (with the exception of flu shots which still widely contain thimerosal), they began making some vaccines using human tissue.

Hospitalization rates higher in kids who get flu shots
2009-05-19, *US News and World Report*
http://health.usnews.com/health-news/managing-your-healthcare/research/articl . . .

Children who get the annual flu vaccine, especially those who have asthma, may be more likely to be hospitalized than children who don't get the shot, a new study shows. But the researcher noted . . . "This may not be a reflection of the vaccine but that these patients are the sickest, and their doctors insist they get a vaccination," said study author Dr. Avni Y. Joshi, a fellow at the Mayo Clinic in Rochester, Minn. "I would be very cautious about interpreting this," said Dr. Gurjit Khurana Hershey, director of asthma research and professor of pediatrics at Cincinnati Children's Hospital. "The bottom line is that kids with asthma who get the flu vaccine are probably a different population anyway. They may be the more severely ill children, so it may have very little to do with the vaccine." The study has too many unknowns and covers too wide an age range over too many flu seasons to indicate any change in recommendations, said Dr. Hank Bernstein, a member of the committee on infectious diseases of the American Academy of Pediatrics. The authors looked back at 263 children aged 6 months to 18 years who had visited the Mayo Clinic between 1999 and 2006 with laboratory-confirmed influenza. **Children—including children who had asthma—who received the annual inactivated flu vaccine were almost three times more likely to be hospitalized than those who were not inoculated.**

Leading doctor: vaccines-autism worth study
2008-05-12, CBS News
http://www.cbsnews.com/stories/2008/05/12/cbsnews_investigates/main4086809.shtml

Jordan King was a typical baby. His parents called him vocal and vivacious. Then just before age 2, after a large battery of vaccinations, he simply withdrew from the world. "The real scary thing was when I noticed he wasn't looking at us any more in the eyes," Mylinda King, Jordan's mother, said. William Mead was a Pottery Barn baby model and met all the typical milestones. Then, also at age 2, after a set of vaccinations, William became very ill and he, too, changed forever. **In both children, batteries of tests revealed dangerous levels of the brain toxin mercury in their systems. Their only known exposure: the mercury preservative once widely used in childhood shots.** Dr. Bernadine Healy is the former head of the National Institutes of Health, and the most well-known medical voice yet to break with her colleagues on the vaccine-autism question. In an exclusive interview with CBS News, Healy said the question is still open. "I think that the public health officials have been too quick to dismiss the hypothesis as irrational," Healy said. Healy goes on to say public health officials have intentionally avoided researching whether subsets of children are "susceptible" to vaccine side effects—afraid the answer will scare the public. **CBS News has learned the government has paid more than 1,300 brain injury claims in vaccine court since 1988, but is not studying those cases or tracking how many of them resulted in autism."**

Did Merck unfairly monopolize the market for a mumps vaccine?
2014-09-10, *Wall Street Journal* blog
http://blogs.wsj.com/pharmalot/2014/09/10/did-merck-unfairly-monopolize-the-m . . .

Did Merck use false pretenses to monopolize the market for mumps vaccines? A pair of lawsuits – one of which is filed by former employees and the other by doctors – make this allegation and a federal judge is allowing both claims to proceed.

The former employees – virologists who filed a whistleblower lawsuit four years ago – charge Merck knew its vaccine was less effective than the purported 95% efficacy level. And they alleged that senior management was aware, complicit and in charge of testing that concealed the actual effectiveness. They claim to have witnessed fIrsthand what they describe as "improper testing and data falsification in which Merck engaged in order to conceal what the drug maker knew about the vaccine's diminished efficacy. In fact, their Merck superiors and senior management pressured them to participate in the fraud and subsequent cover up when they objected to and tried to stop it," according to their lawsuit. The feds declined to join the lawsuit, which was unsealed two years ago. Shortly afterward, the physicians subsequently filed the other lawsuit charge the vaccine was mislabeled and was not the product for which the government or other purchasers paid, which meant that Merck violated the False Claims Act. Both lawsuits note that Merck held an exclusive license to sell a mumps vaccine and its actions discouraged competition. "The ultimate victims here are the millions of children who, every year, are being injected with a mumps vaccine that is not providing them with an adequate level of protection," the lawsuit filed by the virologists states. Meanwhile, the mumps vaccine was ringing the register at Merck, which reported that sales reached $621 million last year.

CDC's vaccine safety research is exposed as flawed and falsified in peer-reviewed scientific journal
2014-06-13, Yahoo! Finance/PR Newswire
http://finance.yahoo.com/news/cdcs-vaccine-safety-research-exposed-115600020 . . .

> Just months after US Congressman Bill Posey compared the Centers for Disease Control (CDC)'s vaccine safety studies to the SEC's Bernie Madoff scandal, malfeasance in the CDC's studies of thimerosal-containing vaccines has, for the first time, been documented in peer-reviewed scientific literature. The journal *BioMed Research International* now provides direct evidence that the CDC's safety assurances about the mercury-containing preservative are not fact-based, according to the article's lead author, Brian Hooker. The paper [cites] over 165 studies that have found thimerosal to be harmful, including 16 studies that had reported [serious detrimental] outcomes in human infants and children. **"Substantial scientific evidence exists and has existed for many years that the vaccine ingredient thimerosal is a developmental neurotoxin"** says George Lucier, former Associate Director of the National Toxicology Program. Studies showing harm from thimerosal sharply contradict published outcomes of six CDC coauthored and sponsored papers – the very studies that CDC relies upon to declare that thimerosal is "safe" for use in infant and maternal vaccines. Dr. Hooker . . . said of the six CDC studies, "Each of these papers is fatally flawed from a statistics standpoint and several of the papers represent issues of scientific malfeasance. For example, **important data showing a**

relationship between thimerosal exposure and autism are withheld from three of the publications. This type of cherry-picking of data by the CDC in order to change the results of important research studies to support flawed and dangerous vaccination policies should not be tolerated."

Family to receive $1.5M+ in first-ever vaccine-autism court award 2010-09-09, CBS News
http://www.cbsnews.com/8301-31727_162-20015982-10391695.html

The first court award in a vaccine-autism claim is a big one. CBS News has learned the family of Hannah Poling will receive more than $1.5 million dollars for her life care, lost earnings, and pain and suffering for the first year alone. In addition to the first year, the family will receive more than $500,000 per year to pay for Hannah's care. Those familiar with the case believe the compensation could easily amount to $20 million over the child's lifetime. Hannah was described as normal, happy and precocious in her first 18 months. Then, in July 2000, she was vaccinated against nine diseases in one doctor's visit: measles, mumps, rubella, polio, varicella, diphtheria, pertussis, tetanus, and Haemophilus influenzae. Afterward, her health declined rapidly. She developed high fevers, stopped eating, didn't respond when spoken to, began showing signs of autism, and began having screaming fits. **In acknowledging Hannah's injuries, the government said vaccines aggravated an unknown mitochondrial disorder Hannah had which didn't "cause" her autism, but "resulted" in it. It's**

unknown how many other children have similar undiagnosed mitochondrial disorder. All other autism "test cases" have been defeated at trial. Approximately 4,800 are awaiting disposition in federal vaccine court.

Gardasil researcher speaks out
2009-08-29, CBS News
http://www.cbsnews.com/8301-500690_162-5253431.html

Amid questions about the safety of the HPV vaccine Gardasil, one of the lead researchers for the Merck drug is speaking out about its risks, benefits and aggressive marketing. Dr. Diane Harper says young girls and their parents should receive more complete warnings before receiving the vaccine to prevent cervical cancer. Dr. Harper helped design and carry out the Phase II and Phase III safety and effectiveness studies to get Gardasil approved, and authored many of the published, scholarly papers about it. She has been a paid speaker and consultant to Merck. **It's highly unusual for a researcher to publicly criticize a medicine or vaccine she helped get approved.** Dr. Harper joins a number of consumer watchdogs, vaccine safety advocates, and parents who question the vaccine's risk-versus-benefit profile. **She says data available for Gardasil shows that . . . there is no data showing that it remains effective beyond five years. This raises questions about the CDC's recommendation that the series of shots be given to girls as young as 11-years old.** "If we vaccinate 11 year olds and the protection doesn't last . . . we've put them at harm from side effects, small but real, for no benefit," says Dr. Harper.

"The benefit to public health is nothing, there is no reduction in cervical cancer, they are just postponed, unless the protection lasts for at least 15 years, and 70% of all sexually active females of all ages are vaccinated." She also says "that enough serious side effects have been reported after Gardasil use that the vaccine could prove riskier than the cervical cancer it purports to prevent. Cervical cancer is usually entirely curable when detected early through normal screenings."

The age of autism: "A pretty big secret"
2005-12-07, *Washington Times*/UPI
https://web.archive.org/web/20051213040915/http://www.washingtontimes.com/upi . . .

Where are the autistic Amish? In Lancaster County, heart of Pennsylvania Dutch country, there should be well over 100 with some form of the disorder. There is evidence of only three. Julia is one of them. She . . . is adopted from China. She had most of her vaccines given to her in the United States. [Of the other, one definitely had a vaccine, and the other's vaccine status is unknown.] Thousands of children cared for by Homefirst Health Services in metropolitan Chicago have at least two things in common with thousands of Amish children in rural Lancaster: They have never been vaccinated. And they don't have autism. **"We have about 30,000 or 35,000 children that we've taken care of over the years, and I don't think we have a single case of autism in children delivered by us who never received vaccines," said Dr. Mayer Eisenstein**, Homefirst's medical director. Eisenstein, in fact, is author of the book "Don't Vaccinate Before

You Educate!" Earlier this year Florida pediatrician Dr. Jeff Bradstreet said there is virtually no autism in home-schooling families who decline to vaccinate for religious reasons – lending credence to Eisenstein's observations. "It's largely non-existent," said Bradstreet, who treats children with autism from around the country. Thimerosal, which is 49.6% ethylmercury by weight, was phased out of most US childhood immunizations beginning in 1999, but the CDC recommends flu shots for pregnant women and last year began recommending them for children 6 to 23 months old. Most of those shots contain thimerosal.

Debate over vaccines, autism won't die
2005-06-26, MSNBC
http://msnbc.msn.com/id/8336821

The afternoon after Kelly Kerns' 2-month-old daughter Kaylee got several vaccines was "living hell," with the child screaming and arching her back, her mother said. 'I kept telling myself everybody gets vaccinated — this is OK,' she said. When Kaylee was 18 months old, her white-blonde hair began falling out and she stopped talking. Meanwhile, Kerns had twin boys — Andrew and Daniel. When they were 15 months old, they received three vaccines. A week later, they stopped talking. All three children have since been diagnosed as autistic. Flu vaccine sold in multidose vials still contains the preservative, and the government urges flu shots for pregnant women and young children even though not enough thimerosal-free ones are available, critics say. Finding answers is tough because autism, a

> little-understood developmental disorder, often is diagnosed at the very ages when children get vaccines. **The stories are remarkably similar: A seemingly normal child gets a shot and days, weeks or months later, withdraws from the world, stops speaking, becomes upset at random stimulation such as a doorbell, and adopts compulsive behaviors like head-banging.**

Note: Thimerosal has now been removed from childhood vaccines. However, the government and drug industry continue to deny that there is any link between mercury in vaccines and autism. And mercury is still commonly used in flu and other vaccines in the US. A document hidden by drug companies for two years all but proves that a commonly used vaccine is responsible for the deaths of countless toddlers within ten days of receiving the vaccine. This document is discussed on the US National Institutes of Health website at this link.

Possible mercury, autism connection found in study
2005-03-17, *Los Angeles Times*
http://www.latimes.com/news/science/la-sci-autism17mar17,1,1770760.story

> Studying individual school districts in Texas, the epidemiologists found that those districts with the highest levels of mercury in the environment also had the highest rates of special education students and autism diagnoses. **There was a strong, direct relationship between mercury and autism levels. The incidence of autism has grown dramatically over the last two decades, from about one in every 2,000 children to as high as one in every 166.** The purported link between autism and mercury has been a subject of intense debate. In the past it has centered primarily on the mercury-containing preservative

> thimerosal, which was once widely used in vaccines. Many parents have argued that thimerosal causes autism because their children seemed to develop the neurological disorder shortly after they received childhood vaccinations.

The man behind the vaccine mystery
2002-12-12, CBS News
http://www.cbsnews.com/stories/2002/12/12/eveningnews/main532886.shtml

> It's been a mystery in Washington for weeks. **Just before President Bush signed the homeland security bill into law an unknown member of Congress inserted a provision into the legislation that blocks lawsuits against the maker of a controversial vaccine preservative called "thimerosal," used in vaccines that are given to children.** Drug giant Eli Lilly and Company makes thimerosal. It's the mercury in the preservative that many parents say causes autism in thousands of children. But nobody in Congress would admit to adding the provision, reports CBS News Correspondent Jim Acosta – until now. House Majority Leader Dick Armey tells CBS News he did it to keep vaccine-makers from going out of business under the weight of mounting lawsuits. "I did it and I'm proud of it," says Armey, R-Texas. "It's a matter of national security," Armey says. Because Armey is retiring at the end of the year, some say the outgoing majority leader is the perfect fall guy to take the heat and shield the White House from embarrassment.

Note: A Reuters article reports that the former head of the US's CDC was later named president of Merck's vaccine division with accompanying high salary. Could this be payoff for her support in suppressing studies that cast doubt on vaccines?

The age of autism: the Amish anomaly
2005-04-18, *Washington Times*
http://www.washtimes.com/upi-breaking/20050321-115921-9566r.htm

> Where are the autistic Amish? Here in Lancaster County, heart of Pennsylvania Dutch country, there should be well over 100 with some form of the disorder. I have come here to find them, but so far my mission has failed, and **the very few I have identified raise some very interesting questions about some widely held views on autism.** The Amish have a religious exemption from vaccination. So far, there is evidence of only three, all of them children, the oldest age 9 or 10. Julia is one of them. She . . . is adopted from China. She had most of her vaccines given to her in the United States before we got her. [Of the other one definitely had a vaccine, and the other's vaccine status is unknown.] The mainstream scientific consensus says autism is a complex genetic disorder, one that has been around for millennia at roughly the same prevalence. That prevalence is now considered to be 1 in every 166 children born in the United States.

Note: The above article appears to have been removed from the *Washington Times* website. You can still find it on the UPI website at this link. Page two is available here. If these links fail, click here.

This article reprinted from http://www.greenmed.com/200-evidence-based-reasons-not-vaccinate-fr . . . It provides good links to over three hundred documents regarding adverse health effects due to vaccinations.

200 Evidence-Based Reasons NOT to Vaccinate—FREE Research PDF Download!

The media, your pediatrician, politicians and health authorities like the CDC and FDA claim that vaccines are safe and effective. So why do hundreds of peer-reviewed studies indicate the opposite is true? Read, download, and share this document widely to provide the necessary evidence-based counterbalance to the pro-vaccination propaganda that has globally infected popular consciousness and discussion like an intractable disease.

It is abundantly clear that if the present-day vaccine climate, namely, **that everyone must comply with the CDC's one-size-fits-all vaccination schedule or be labeled a health risk to society at large**, is to succumb to open and balanced discussion, it is the peer-reviewed biomedical evidence itself that is going to pave the way toward making rational debate on the subject happen.

With this aim in mind, GreenMedInfo.com has painstakingly collected over 300 pages of study abstracts culled directly from the National Library of Medicine's **pubmed.gov** bibliographic database on the wide-ranging adverse health effects linked to vaccines in today's schedule (**over 200 distinct adverse effects, including death**), as well as numerous studies related to vaccine contamination, and vaccine failure in highly vaccine compliant populations.

This is the literature that the media, politicians and governmental health organizations like the CDC, pretend with abject dishonesty does not exist – as if vaccine injury did not happen, despite the over 3 billion dollars our government has paid out to vaccine injured through the National Vaccine Injury Compensation Fund since it was inaugurated in 1986.

We have written extensively about this research previously, highlighting different studies, focusing on translating their implications to the lay persons (**view our vaccine article section here**), but we believe that collecting and condensing solely the primary literature itself makes a much more powerful statement.

This document is **being made free to download** to the world at large in order to encourage the lay public, health professionals, activists, and elected officials alike to read, acknowledge and share the voluminous literature with their family, friends, colleagues and related stakeholders. You will find this research undermines the national and global agenda to continue to expand the vaccine schedule (on behalf of a vaccine industry that is indemnified against lawsuit for defective or harmful products), with increasing legislative pressure to remove exemptions and mandate them against the evidence of harm and at best equivocal effectiveness as a preventive health measure.

If the vaccination arm of modern medicine today is to continue to promote itself as a science- and evidence-based practice, *it must* acknowledge and incorporate the implications of the research we are releasing here, or lose any pretense at credibility. Failing to do so will reveal that the widespread push to remove your choice in the matter is agenda and not evidence driven, and due to the fact that vaccines all carry the risk of **irreversible harm and even death** (any vaccine insert proves this), it clearly violates the Nuremberg code of medical ethics to promote them as *a priori safe* and effectiveness. **Also, please sign the Whitehouse petition: "Prohibit Any Laws Mandating the Force and Requirement of Vaccinations of Any Kind."**

Graphical Representation of Historical Diseases

Measles mortality graphs are enlightening [more below] and contradict the claims of Government health officials that vaccines have saved millions of lives.

It is an unscientific claim which the data show is untrue. Here you will also learn why vaccinations like mumps and rubella for children are medically unethical and can expose medical professionals to liability for criminal proceedings and civil damages for administering them.

Typhoid and Scarlet Fever – Mortality UK, USA and Australia, Typhoid and Scarlet Fever vanished without vaccines but with clean water, better nutrition, sanitation and living conditions.

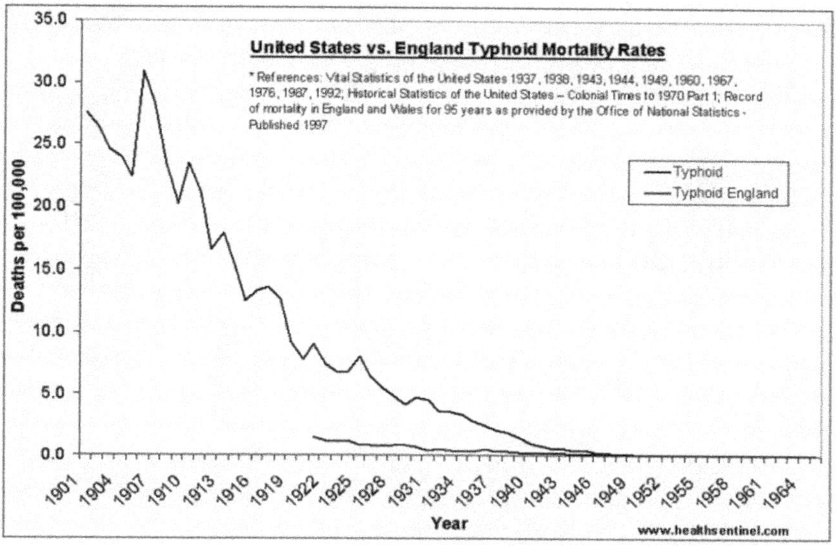

The graph below is from a peer refereed medical paper by Englehandt, S. F., Halsey, N. A., Eddins, D. L., Hinman, A. R., "Measles mortality in the United States, 1971–1975. *Am J Public Health* 1980; 70:1166–1169. The red dotted trend line has been added.

This shows US measles mortality was falling regardless of whether vaccination was used. By 2010, overall measles mortality in the USA was to fall to around 1 in 25 million without vaccines. As the severity of measles declined, long-term complications would also. Whilst people still caught measles, it was not the dreaded disease we are told it is today.

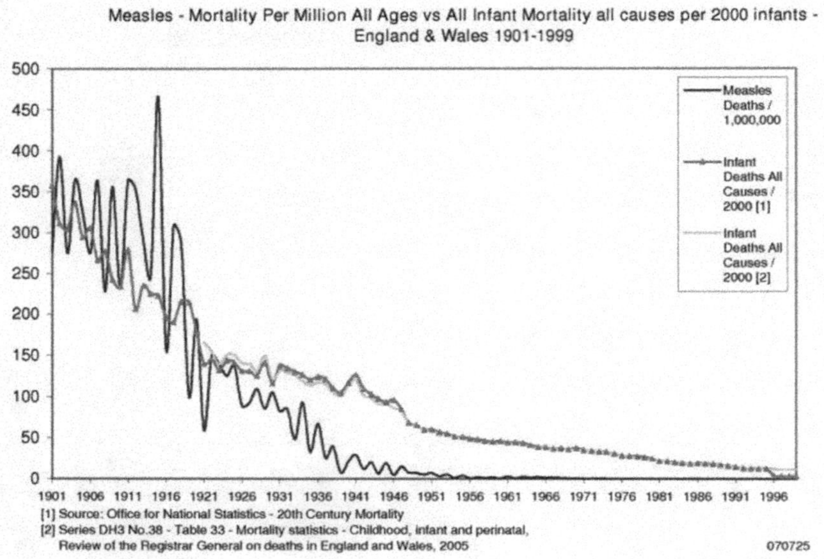

Mortality, life expectancy, healthcare costs UK, USA, and worldwide

Does paying for healthcare bring you better health and a longer life? No.

The following graphs show that in 1996, average life expectancy in the US was eighteenth of all countries, being 5 years less than Canada and behind the UK. But Americans were paying per person US $1000 or over 1/3rd more than Canadians and nearly 2/3rds more than the British. And if you then take a look at the graphs of mortality, what were Americans getting for their money? Mortality rates were falling anyway, regardless, and kept on falling. Life expectancy increased as time went by, but again substantially due to overall improved living conditions.

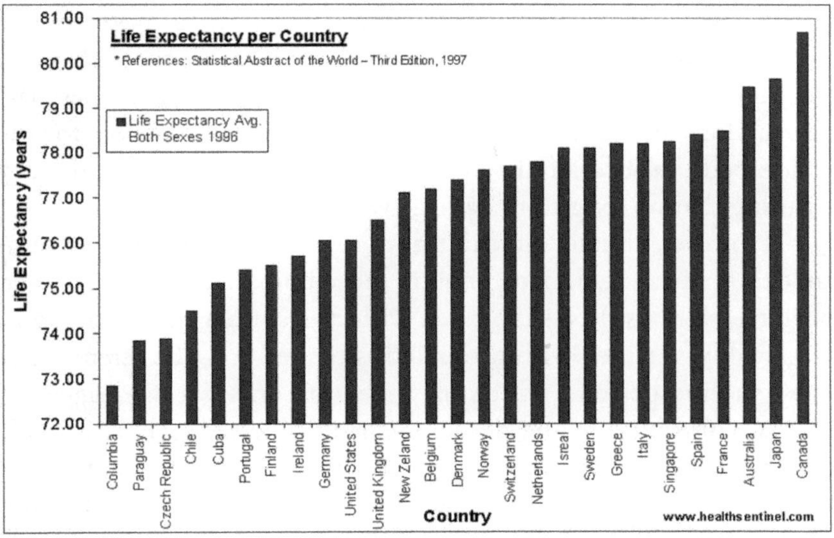

Why vaccines are harmful

The United States legal standard applied to vaccines defines them as "unavoidably unsafe products that are quite incapable of being made safe for their intended and ordinary use." The reason why vaccines are unsafe, or in other words harmful, is because they are made up of chemicals and other elements that are poisonous to the body. Some of these chemicals and elements include formaldehyde, which is commonly used to embalm corpses and is a known carcinogen (capable of causing cancer), thimerosal, a derivative of mercury which is a toxic heavy metal, and aluminum phosphate which is a toxin used in deodorants. Other toxic ingredients include phenol (carbolic acid), alum (a preservative), and acetone which is a volatile solvent used in fingernail polish remover.

Vaccines contain not only poisonous chemicals but also foreign proteins such as chick embryo, calf serum, rabbit brain tissue, and monkey kidney cells. These foreign proteins can trigger numerous allergic and inflammatory reactions and can produce anaphylactic shock in susceptible infants. When these vaccine ingredients enter the child's bloodstream (either through injection or taken orally), the child's body will take immediate action to eliminate these poisons either through the normal organs of elimination or through acute reactions such as fever, swelling or skin rashes. As long as these latter reactions are not suppressed through drugs, it is possible for the child's body to successfully eliminate these vaccine poisons, thereby ensuring that no long-term damage will occur.

However, should the child have insufficient vitality to mount an eliminative reaction or should the eliminative reaction (fever, swelling, skin rash) be suppressed with drugs, then the vaccine poisons will be retained within the body's tissues. It is the retention of these vaccine poisons, which in susceptible infants, can trigger or at least contribute to the development of illnesses such as juvenile diabetes, autism, asthma, neurological disorders, leukemia, and even cot deaths. In many children, the retention of vaccine poisons within the body may not cause any acute or noticeable symptoms, but it will cause a lowering of the child's vitality, which in turn, weakens and impairs its intellectual, creative, and imaginative powers, its physical energy and strength, and most all of its internal metabolic functions and immune activities. What this means is that the child will operate at a level well below its true potential.

The toxic buildup within the child's body and the subsequent depletion of its vitality through vaccines, drugs, fluoridated water, food additives, pesticide residues, atmospheric pollutants, electromagnetic radiation and other adverse factors in the lifestyle makes the child more susceptible to chronic disease at an earlier age, and it's worth noting that the dramatic rise in childhood cancer, autism, juvenile diabetes, asthma, and neurological disorders over the past fifty years has directly coincided with the era of drugs, vaccines and chemical poisons in our food, water, and environment.

(Further reading: Dr. Mark Randall, vaccine whistleblower; "What you do when you vaccinate," Dr. Stanley Bass; immunization studies: scientific and medical references; vaccines and immune suppression; Janine Roberts website; Whale's vaccine website)

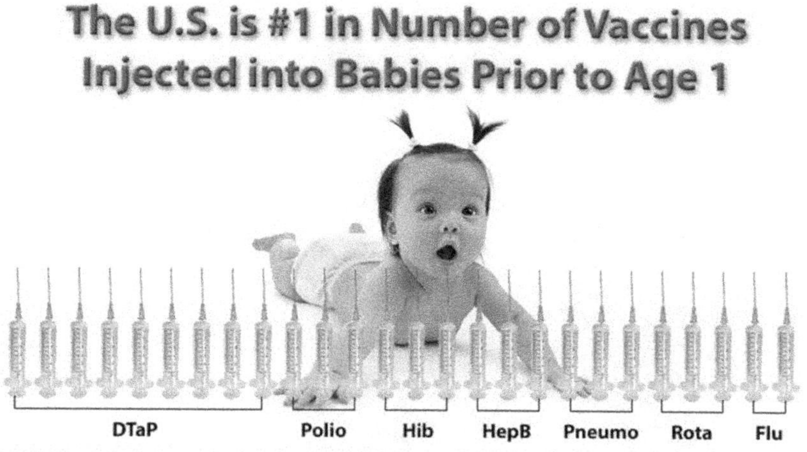

Further reading: *Vaccines and Disease*; *Vaccination: Assault on the Species*; *The Polio Vaccine Myth*

The below graphs, based on the official death numbers as recorded in the official year books of the Commonwealth of Australia, are taken from Greg Beattie's excellent book *Vaccination: A Parent's Dilemma* and represent the decline in death rates from infectious disease in Australia. They clearly show that vaccines had nothing to do with the decline in death rates. (Note: Graphical evidence on the decline in death rates from infectious disease for USA, England, New Zealand, and many other countries shows the exact same scenario as above). So what were the true reasons for this decline? From his book *Health and Healing*, Dr. Andrew Weil best answers it with this statement: "Scientific medicine has taken credit it does not deserve for some advances in health. Most people believe that victory over the infectious diseases of the last century came with the invention of immunizations. In fact, cholera, typhoid, tetanus, diphtheria and whooping cough, etc., were in decline before vaccines for them became available—the result of better methods of sanitation, sewage disposal, and distribution of food and water."

Measles

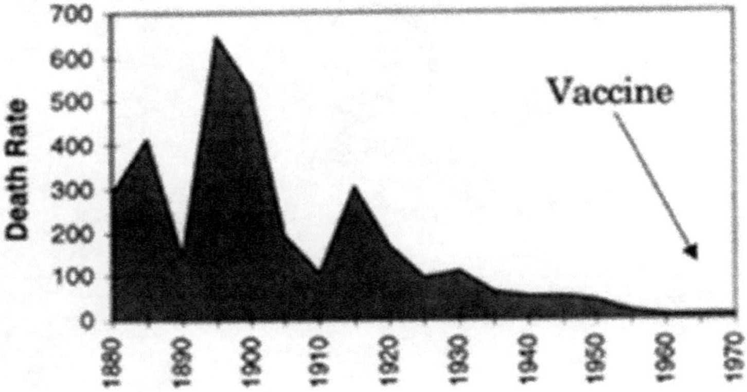

Scarlet Fever

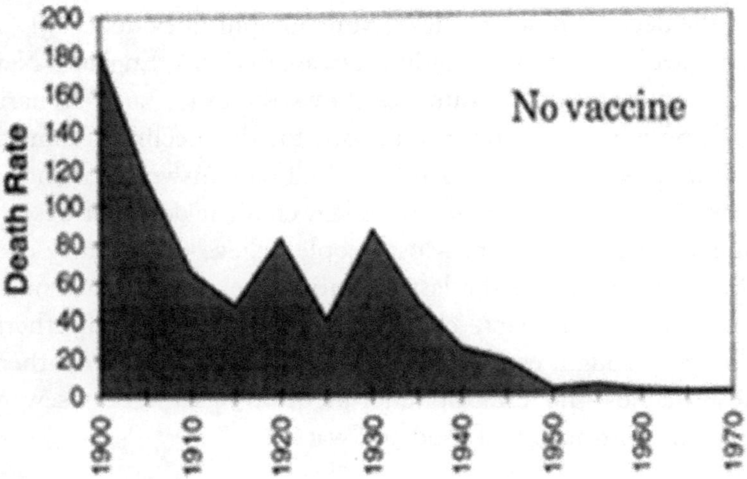

Chapter 15

Conclusion

In my Internet search for the relevant information regarding the few "silent killers" covered in this book, I was overwhelmed by the amount of information available. There are hundreds, upon hundreds of studies indicating harmful health effects. The few toxins that were examined in this book are just a drop in the bucket of the numerous chemicals that the human body is plagued with. Reports indicate that the newborn babies already contain over two hundred different chemicals in their bodies at birth. I have also provided the reader with numerous Internet links regarding this subject matter so that you can verify this information for yourself and make your own conclusion.

Is it any wonder that the diseases afflicting mankind are on an increase? All these toxins act synergistically, reinforcing their toxic effect on our health, disrupting the endocrine system and our gut flora, the first important line of defense to our health. That is why the diseases are on an increase due to the poisons in our water, food, and air. There are also reports that infertility is on the rise also. There can only be one answer to this—we are slowly being poisoned—that is the conclusion I came to. You, dear reader, will have to make your own conclusion.

Our health system has come up with numerous medications to alleviate our health problems, but all we get are synthetic medications that mask the symptoms of a disease. **This is called a managed**

health care system to ensure one keeps going back to the health provider. It ensures a good income for the health provider, a good income to the pharmaceutical companies, and a financial drain on the consumer. That is why they do not want any natural cures since these would cure the patient. That is why they go into a frenzy discrediting natural cures so they can propagate their poisonous medicines. These medications do not cure any disease. When one looks at the inserts, there are two pages of side effects that may require additional synthetic medication to mask all the adverse health effects. And how about all the law firms advertising regarding some of these medications that a consumer can file a lawsuit against? Television ads are full of them. Yet the controlled news media does not even mention the millions of dollars the pharmaceutical companies have to pay in settlements. Below is a graph of deaths attributed to the approved drugs by FDA. The source for this graph was published by FDA:

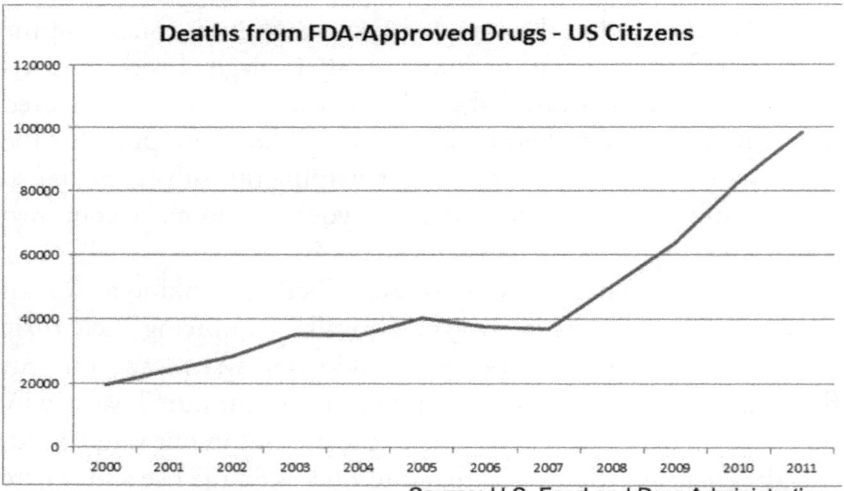

Source: U.S. Food and Drug Administration

To make matters worse, our health watchdog federal agencies have undergone deteriorated transformation from health conscious organizations, to rubber-stamping whatever poisons biotech industries or pharmaceutical companies submit for approval. An astronomical change has occurred to the FDA (Federal Food and Drug Agency) since I was employed by them. The agencies are overrun by/with biotech executives, watching out for industry profits, or maybe even a more sinister assignment. It is also true in other federal agencies like EPA, USDA, etc. They are watching out for the industries wellbeing instead for our health of the American citizens. Don't get me wrong. There are a lot of lower-echelon conscious scientists at these organizations, but it is the higher level of executives that has infiltrated from the biotech companies who make the decisions that harm our health.

I was employed by the FDA in the fall of 1961 upon my graduation from Wayne State University with a bachelor of science degree in analytical chemistry. The FDA Detroit District in 1961 employed over twenty bench chemists and a large contingent of inspectors. The main assignment of each FDA District in those times was to safeguard the health of the American consumer from fraudulent manufacturing practices. The inspectors would go out into the field and check on manufacturing industries in food and drugs. They would observe sanitary procedures that had to be strictly adhered to, used in the manufacturing process, and collect samples for analysis.

As an example, if a manufacturer listed food content in a can or package, he was allowed a certain minimum or maximum standard deviation from the content amount listed on the label. The labels were also required to list all the ingredients, and in the case of drugs, the amount. If a manufacturer of canned food listed 8.5 oz., he was allowed $\pm$0.25 oz. deviation. Suppose the manufacturer was producing eight thousand cans per day and was only putting 8.0 oz. per can, he was able to produce an additional five hundred cans at no cost to him for contents, but a profit at the expense of cheating the consumer.

Since the middle of the 1970s, the workforce has been gradually decreased or completely eliminated in most districts. The safety of food and drugs was left up to the manufacturing industry with a requirement of submission of any safety studies. **This is like asking the fox to guard the hen house.** Is it any wonder that toxic substances such as aspartame, fluoride, MSG, artificial coloring, or taste flavors have been introduced into our food supply? FDA, EPA, and USDA are relying on studies submitted by the manufacturers or joint studies in collaboration with private companies or academia. Can one believe in the reliability of these studies? The manufacturers foot the bill for these studies. If they don't get their favorable results, do you think they will ask these outside companies or academia for additional studies? The independent studies indicate a highly different picture as to extreme adverse health effects on our well-being as was indicated under each individual toxin discussion. Remember reading the aspartame approval process? For a complete history of aspartame approval access the link provided. There is a massive increase in numerous diseases as was evidenced under the individual toxins discussed previously. This probably is due to many executives and managers from biotech and food industry having transitioned themselves into prominent, authoritative positions within the federal agencies. There are numerous reports of such persons joining FDA, EPA, USDA, CDC, and NIH who made or are making decisions to benefit the industry as opposed to the safety of the American consumer. This was evidenced in the approval process of the toxin aspartame. How about the deceit in water fluoridation?

http://www.globalresearch.ca/ . . . /5336422

http://progressivecynic.files.wordpress.com/2012/11/tumblrmlmsl1mwcp1r6m2leo1_500.jpg

http://progressivecynic.files.wordpress.com/2012/11/geke.png

Any independent studies submitted to these organizations that have been infiltrated by industry executives are usually met with a statement "We need more studies." Who are they kidding? There are hundreds, and hundreds of highly reliable independent studies verifying the toxic effect on human health but are **completely ignored by the pseudo regulators infested in our federal agencies.**

This raises a very important question. Where are the environmentalists? They raise a lot of furor and public awareness regarding some insect, bird, or animal about possible extinction due to habitat destruction. How about the human race? All these toxins are taking a toll on human life. The proof is in the pudding with all the diseases that are on the rise, like cancer, heart disease, obesity leading to diabetes, gastrointestinal diseases, and an alarming rate increase in autism. The only conclusion that I came to, is that we are slowly being poisoned, but you, dear reader, will have to draw your own conclusion, after accessing the additional information from the Internet links provided.

The conclusion that I have come to is that men are more inclined toward evil than good. Greed is the number one goal of most corporations under the disguise of profits for shareholders, to the point of eliminating the competition via any means. Prime example is Big Pharma, which does not allow the medical profession to use alternative holistic type treatment due to natural cures. The medical schools cannot teach alternative medicine. Doctors who practice alternative medicine using natural means are threatened with loss of their medical license by the AMA. Pertinent and highly effective natural cures are suppressed by the mainstream medical profession. Prime example is the Royal Rife cancer cure method invented over eighty years ago. Monsanto bought up most of seed companies to control most of the agricultural arena. To add insult to injury, for any cross-pollination resulting from their GMO seeds to neighboring farms, these farmers face a lawsuit for patent infringement. It should be the other way around, due to contamination, and in some instances loss of revenue because of the contamination of product. Due to some unethical politician who introduced a rider to a bill and passed by Congress, Monsanto cannot be sued.

Preach liberalism. I think most people are aware of this presently. Anyone who does not understand this trend must be mentally deficient due to the fluoride and aspartame among others, having an effect on their brain. How about the chemtrails and vaccines with the aluminum and mercury content that have an adverse effect on the mental health.

Use the idea of class warfare, with plenty of examples by the Obama administration, especially by the Department of Justice. Using any means to achieve their goals is pretty well self-explanatory. How about this one—to lie in force. When was this Obama administration truthful? "If you like your doctor, you can keep your present doctor!" How about the tax cover-up, NSA spying on the citizens, and the Benghazi cover-up. And the list goes on and on. This was to be a more transparent form of government, but what we received is a more secretive and corrupt government than any other administration in the history of our country. And the mainstream news media completely ignores all these instances. Yet any criticism of the Obama administration by anyone, the mainstream news media attacks them unmercifully trying to discredit that person or organization. This reminds me of a newspaper called *Pravda*, meaning *Truth*, which was published by the communist regime in Russia. Except it was far from the truth. It was more of a propaganda type media.

Placing regulatory obstacles in our industry with numerous expensive regulations. Use alcohol, drugs, corruption, and all forms of vice to systematically corrupt our youth by any means. The education system is changed with a new system like Common Core, which changes the history of our nation, further confusing our youth.

Political candidates chosen and placed in public office are more obedient to industry demands and will be used as pawns in the name by those behind the scenes. There are no public records on president Obama. Was he chosen some time ago and groomed to become the president of the United States, to ensure the decadence that has been created in our country over the past few years? To increase our debt and to saddle our industry with highly restrictive, expensive, and prohibitive regulations? To use high taxes and unfair competition to bring economic control as well as economic chaos?

They use United Nations laws to destroy the sovereignty of the United States and enslave the people. This will be accomplished via gun control and Internet control to eliminate a lot of pertinent and useful information to keep the population dumb and misinformed. They use toxic chemicals in the air such as chemtrails, fluoride in the water, and a number of toxic food additives, genetically modified crops containing pesticides, herbicides, and insecticides, and vaccinations. They tell our federal health agencies that if serious questions are raised regarding safety, to stall and request more studies. They accomplished this by having these agencies such as FDA, EPA, USDA, AMA, ADA, NIH, etc., infiltrated at the highest level by individuals loyal to the globalist administration.

So let's ask ourselves a few questions. Is this the pattern taking place in America? Are we facing a massive financial collapse due to the increasing debt? The potential mass extermination of the population due to the use of toxic chemicals in the air, water, and food, compounded by the increased use of pesticides, insecticides, and herbicides that are carcinogenic not only to humans but our ecology like the beneficial bacteria in the soil and detrimental to our pollinators like the bees and the monarch butterfly. And now there is talk about mandatory vaccinations, where it already has been verified that sterilizers have been used worldwide, and an increase in other toxic elements in them. All this with a slogan—vaccines are good for you, and the dumbed-down individuals believe it. What a novel idea to speed the demise of the civilization as we know it.

Most European countries have much more stringent laws. They have banned water fluoridation, azodicarbonamide in their baked goods, question and almost stopped GMO crops. The only toxic chemicals applied worldwide are chemtrails, aspartame, and vaccinations, plus the use of herbicides, pesticides, and insecticides. Borrowing a statement from the Shield Report under chemtrails, the survival of this planet for only about thirty to fifty years can become reality if we continue to poison our air, food, and water. It is not due to the radiation of ultraviolet rays but to humans poisoning themselves and destroying the ecosystem as well.

Another very interesting article was written by Mr. Dean Garrison, an editor and writer for DCClothsline.com. It is reprinted with permission:

> Recently, we have posted a couple of articles at DCClothsline that question the narrative of the current Ebola outbreak. DCClothsline is not just news site but a site that explores the world conspiracy and embraces the truth movement. The problem with looking at conspiracy is that often times people immediately disregard them. You can tell people things like 'please just look at this video', or 'please just read this article,' and it

really does not matter. People are busy and your conspiracy theory is not their main priority. If they are not busy they are simply just too lazy or closed-minded.

So today, rather than tell you why you should believe anything about a possible Ebola conspiracy, I just want to show you some quotes from very famous people both past and present. If the words of these people do not prove to you that the elite of the world are looking for a way to reduce our population, then nothing likely will. Before you can believe that Ebola, or any other virus might be used in such a way, you must first understand that there are a lot of people out there who would like billions of the world's inhabitants dead.

Ever heard of a guy by the name of Bill Gates? Bill Gates is what many call a eugenist, but he is not alone and is surrounded by lots of famous people and organizations worldwide.

'If I were reincarnated I would wish to be returned to earth as a killer virus to lower human population levels.'—Prince Phillip, Duke of Edinburgh, Leader of the World Wildlife Fund.

'Since the inception, the UN has advanced a worldwide program of population control, scientific human breeding, and Darwinism.'—Claire Chambers, The SIECUSCircle; A Humanist Revolution.

'The present vast overpopulation, now far beyond the world carrying capacity, cannot be answered by future reductions in the birth rate due to contraception, sterilization, and abortion, but must be met in the present by reduction of numbers presently existing. This must be done by whatever means necessary.'—Eco-92 Earth Charter

'The United Nation's goal is to reduce population selectively by encouraging abortion, forced sterilization, and control of human reproduction, and regards two-thirds of the human population as excess baggage, with 350,000 people to be eliminated per day.'—Jacque Cousteau

'Childbearing (should be) a punishable crime against society, unless the parents hold a government license . . . All potential parents (should be) required to use contraceptive chemical, the government issuing antidotes to citizens chosen for childbearing.'—David Browser, Executive Director of the Sierra Club.

'If radical environmentalists were to invent a disease to bring human populations back to sanity, it would probably be something like AIDS . . . It has the potential to end industrialization, which is the main force behind environmental crisis.'—Earth First

'At present the population of the world is increasing at about 58,000 per diem. War, so far, has had no very great effect on this increase, which continued throughout each of the world wars . . . War was hitherto been disappointing in this respect, but perhaps bacteriological war may prove effective.'

'If a Black Death could spread throughout the world once in every generation, survivors could procreate freely without making the world too full. The state of affairs might be unpleasant, but what of it?'—Bertrand Russell, The Impact of Science on Society

'A total of population of 250-300 million people, a 95% decline from present levels, would be ideal.'—Ted Turner, CNN founder and supporter of UN

'There are many ways to make the death rate increase.'—Robert McNamara, New Solidarity, March 30, 1981

'The resultant ideal sustainability population is hence more than 500 million but less than a billion.'—Club of Rome, Goals for Mankind

'In order to stabilize world population, we must eliminate 350,000 people per day. It is a horrible thing to say, but it is just as bad not to say it.'—Jacque Cousteau, 1991, explorer and UNESCO courier

'I believe that human overpopulation is the fundamental problem on Earth Today (and) We humans have become a disease, the Humanpox.'—Dave Forman, Sierra Club, cofounder of Earth First.

'We don't want the word to go out that we want to exterminate the Negro population.' —Margaret Sanger, Founder of Planned Parenthood.

'Society has no business to permit degenerates to reproduce their kind.'—Theodore Roosevelt

'An advanced forms of biological warfare that can "target" specific genotypes may transform biological warfare from the realm of terror to a politically useful tool.'—The project for a New American Century, Rebuilding American's Defense, p. 60, Dick Cheney and Paul Wolfowitz.

'Whatever the price of the Chinese Revolution, it has obviously succeeded not only in producing more efficient and dedicated administration, but also in fostering high morale and community purpose. The social experiment in China under Chairman Mao's leadership is one of the most important and successful in human history.'—David Rockefeller, Banker, Honorary Director of Council on Foreign Relations, honorary chairman of Bilderberger Group, Founder

of Trilateral Commission, Member of Bohemian Club, praising Chairman Mao, whose policies killed 30 million people.

'War and famine would not do. Instead, disease offered the most efficient and fastest way to kill billions that must soon die if the population crisis is to be solved. AIDS is not an efficient killer because it is too slow. My favorite candidate for eliminating 90% of the world's population is airborne Ebola (Ebola Reston), because it is both highly lethal and it kills in days, instead of years. "We've got airborne diseases with 90% mortality in humans. Killing humans. Think about that. You know, the bird flu's good, too. For everyone who survives, he will have to bury nine.'—Dr. Eric Pianka

'The world today has 6.8 billion people . . . that's headed up to 9 billion. Now if we do a really great job on new vaccines, health care, reproductive health services, we could lower that by perhaps 10 to 15 percent.'—Bill Gates statement at TED Conference, October 1, 2010.

If you throw these people and organizations into a mix, you might understand that there really are those who are evil enough to try to exterminate a large part of the world populations. Is that what **Ebola** is for. I don't know. I am simply a truth seeker who likes to look at all angles. But if you don't realize that there are people out there who want to exterminate 85% (or more) of the world's population then you might not even bother to look. You might dismiss it all as conspiracy.

"**Agenda 21**" is real. It is dismissed in many clever ways such as **women's rights and environmental responsibility.** In the end, it is really about a bunch of elites who do not believe in

God and have chosen to try to take His place. They think they know what is best for you. You belong to a subspecies. And most of that subspecies needs to be eliminated. That is how these people think. This is the truth, just so you know, I could have listed hundreds more. We really aren't the crazed conspiracy theorists that you might think, at least not all of us. Many of us, like myself have just decided to do research that most are unwilling to do."

Meet John Holdren, Barack Obama's Science Advisor

If you think this administration is in favor of population control, then maybe Holdren's words will convince you.

'Indeed, it has been concluded that compulsory population-controlled laws, even including laws requiring compulsory abortions, could be sustained under existing Constitution if the population crisis became sufficiently severe to endanger the society.'—John P. Holdren, Obama's science advisor, ***Ecoscience****, 1977*

'Adding a sterilant to drinking water or staple foods is a suggestion that seems to horrify more than most proposals for involuntary fertility control.'—John P. Holdren, Obamas science advisor, ***Ecosciencw****, 1977*

'Perhaps those agencies combined with UNEP and the United Nations population agencies, might eventually be developed into a Planetary Regime sort of an international superagency for population, resources, and environment. Such a comprehensive Planetary Regime could control the development, administration, conservation, and distribution of all natural resources, renewable or nonrenewable, at least as insofar as international implications exist. Thus the Regime could have the power to control pollution not only in the atmosphere and oceans, but also in such freshwater bodies as rivers and lakes that cross international boundaries or that discharge into the oceans. The Regime might also be a logical central agency for regulating all international trade, perhaps including assistance from DC'a and LDC's and including all food on

the international market.' John P. Holdren, Obama;s science advisor, ***Ecoscience****, 1977.*

Sources: http://freedomoutpost.com/2014/08/us-not-a-conspiracy=theory-when-elitists-are-on-record-for-95-of-the-population-to-be-elimited/#IYCTZ7770eRq49Bx99

When one considers what is taking place here in America and to a lesser extent worldwide, it makes one think what is happening to the civilization as we know it. Some of the items mentioned in the article by Mr. Dean Garrison have already occurred, namely use of vaccinations that contained a sterilizer. The minister of health from Kenya reported that the tetanus vaccine contained a sterilizer that was given to girls of child bearing age. Supposedly this vaccine was tried sometime in the mid-1990s in Mexico and Philippines. A little closer at home, the most recent flu vaccine was declared not effective by CDC. Yet CDC was recommending to still take this vaccine even though an independent lab analysis stated it contained 25,000 ppm of mercury. Among other adjuvants that the vaccines contain were aluminum and MSG, potent neurotoxins. The most recent Ebola outbreak in the West African countries could that also be a test of its lethal effectiveness? No proponents want to establish mandatory vaccinations since history of which has not established any health benefits but could potentially be used for mass extermination. It would be another weapon of mass destruction added to the ones already in existence.

The Destruction of America

One big question is, how does one destroy a country like the United States of America? One cannot bring it down militarily, since that is an impossibility due to its armed citizenry. The only answer has to come from within. What methods, then, are available to accomplish this task? One is through education, but the schools teach the history and the liberty guaranteed by the Constitution. But the educational system since has been infiltrated by liberal thinking academics. Then there are a number of federal agencies that safeguard the American liberties and health. These too have been infiltrated by personae who hold allegiance to the industry. One such agency, although not under federal ownership is the AMA (American Medical Association), jointly with Big Pharma, which has established itself as a dictatorial agency, that really has an absolute control over the medical profession. This came about in the nineteen thirties when they were able to suppress the cancer cure that was invented by Royal Rife.

He discovered an incredibly simple, electronic approach to curing literally every disease on the planet caused by viruses and bacteria. Unfortunately, he was branded a medical heretic. Medical history has numerous stories of genius betrayed by backward thought and jealousy but most pathetically greed and money. Semmelweis struggled to convince surgeons to sterilize their instruments and use sterile surgical procedures. Pasteur was ridiculed for his theory that germs could cause disease. Scores of other visionaries went through hell for simply challenging the medical profession on their status quo. Such legendary were scientists and doctors like Roentgen and his x-ray machine, Morton for promoting anesthesia, and Harvey for his theory of circulation of blood. In recent years, these are scientists and doctors such as Dr. W. F. Koch and his molecular therapy, Dr. Emanuel Revici and his nontoxic, chemotherapy approach in curing cancer, Dr. S. Burzynski and his antineoplaston use against cancer, and Dr. R. Hoxsey and his alternative cancer treatment. To this list you can add Dr. Phyllis Mullenix, who was accused of playing politics, and all she wanted to do was to prove the toxic effect of fluoride on the brain cells. There are many more MDs and ODs who have been persecuted and threatened with their medical license suspension so they could not practice medicine. Orthodox big-money medicine resents and seeks to neutralize and/or destroy those who challenge its beliefs. The elimination of number of doctors who do not adhere to mainstream medical practice as was evidenced in June/July in 2015. They died under mysterious circumstances and classified as suicide. What follows is exactly such a sensational therapy and what happened to it.

Royal Raymond Rife, developed technology which is still commonly used today in the fields of optics, electronics, radiochemistry, biochemistry, ballistics, and aviation. Rife developed bioelectric medicine. Rife spent sixty-six years designing and building medical instruments. He mastered many different skills and knowledge of different scientific fields. Whenever new technology was needed to perform a new task, he simply invented and built it himself. By 1920, Rife had finished building the world's first virus microscope. He perfected it, and in 1933, he built the complex universal microscope, containing six thousand parts and able to magnify objects sixty thousand times their normal size. He was the first human being to see a live virus. Modern electronic microscopes kill any living organism beneath them.

Rife identified the individual spectroscopic signature of each microbe by the use of his slit spectroscope attachment. He rotated a block quartz prism to focus light of a single wavelength upon the microbe that was being examined. A wavelength was selected when the spectroscopic signature frequency resonated with the frequency of the microbe. Rife reasoned that every living microorganism has a resonating frequency, and if one could amplify that frequency, it would kill the microorganism. The other theory he had was that no two microorganisms had the same resonating frequency. The result of using a resonant wavelength is that microorganisms, which are invisible in white light, suddenly become visible in a brilliant flash of light when they are exposed to the color frequency that resonates with their own distinct spectroscopic signature. Rife was able to see these invisible organisms and watch them actively invade tissue cultures that no one else could see under the ordinary microscope.

Royal Rife identified the human cancer virus first in 1920. He named the cancer virus *Cryptocides primordiales*. Rife was also able to view virus changes into different forms under his microscope. At the time, it was condemned by other influential doctors who did not look through Rife's microscope. Nothing can convince a closed mind. Many scientists and doctors have confirmed Rife's discovery of the cancer virus and its pleomorphic nature. Rife used the technique of what made the organism visible of the resonating frequency by increasing the intensity to a point that it killed the organism. Increasing the natural organism's oscillations to such a degree until they distorted and disintegrated the organism from structural stresses. Rife called this the mortal oscillatory rate (MOR), and it did not harm the surrounding tissue. This principle is illustrated by using an intense musical note that can shatter a wine glass. Because everything else has a different resonant frequency, nothing but the glass is destroyed.

It took Rife many years to discover the frequencies which specifically destroyed herpes, polio, spinal meningitis, tetanus, influenza, and number of other dangerous organisms. In 1934, the University of Southern California provided Royal Rife with terminally ill cancer patients. A research committee was established consisting of doctors and pathologist to examine the patients. After ninety days of treatment, the committee concluded that 86.5% of the patients had been completely cured. The 13.5% patients also responded within the next four weeks of treatment. The total recovery rate using Rifes' technology was 100%.

On November 20, 1931, forty-four of the most respected medical authorities honored Royal Rife with a banquet billed as "The End to All Diseases" at the Pasadena Estate of Dr. Milbank Johnson. But by 1939, almost all of these distinguished doctors and scientists were denying that they had ever met Rife, reminding one of Judas. This happened because of one individual—Morris Fishbein, the head of AMA, who threatened doctors with loss of their medical practice. Tyranny was born in America.

A token attempt was made to buy Royal Rife's technology by Morris Fishbein, who acquired the entire stock of the American Medical Association (AMA) by 1934. Rife refused. Morris Fishbein had a history of persecuting scientists or medical doctors who did not cooperate with him. As an example, Dr. Harry Hoxsey, who came up with alternate herbal cancer treatment. Morris Fishbein and his associates presented the following proposal to Dr. Hoxsey: Fishbein's associates would receive all profits for nine years and Hoxsey would receive nothing. Then if they were satisfied, Hoxsey would receive 10% of the profits starting in the tenth year. Talk about mafia style of conducting business. When Hoxsey turned them down, Fishbein used his immensely powerful political connections to have Hoxsey arrested 125 times in a period of sixteen months. The charges were always thrown out of court, but the harassment drove Hoxsey insane. Needless to say, Fishbein could not use the same tactics against Royal Rife. A trial on trumped up charges against Royal Rife would provide the defense an opportunity to introduce evidence such as the 1934 medical study regarding the 100% cure of terminally ill cancer patients. On top of that, Rife had spent decades accumulating meticulous evidence of his work, including film and stop-motion picture photographs. Different tactics had to be used.

The first incident was the gradual stealing of components, photographs, film, and written records from Royal Rife's laboratory. Then someone vandalized Rife's precious universal microscope by stealing certain pieces. At the same time, a multimillion-dollar Burnett Lab in New Jersey was destroyed by an arson fire. The lab was in process of confirming Rife's work. The final blow came when police illegally confiscated the remainder of Rife's fifty years of research. Beam Ray Corporation, which was the only company producing Rife's frequency instrument, was hit with a frivolous lawsuit. The assisted legal assault bankrupted the company due to legal expenses, resulting that the commercial production of Rife's frequency instruments ceased completely. Lawyers followed almighty dollar.

Doctors who tried to defend Rife lost their foundation grants and hospital privileges. They were also threatened with a loss of their medical practice license. Money was also spent on doctors who saw Rife's therapy to forget what they saw. As an example, Dr. Arthur Kendall, director of the Northwestern School of Medicine, who worked with Rife on the cancer virus, accepted almost a quarter of a million dollars to suddenly retire in Mexico. Between the carrots and the sticks, everyone except Dr. Couche and Dr. Milbank Johnson, gave up Rife's work and went back to prescribing drugs. (http://www.dfe.net/Milbank_Johnson.html). To completely finish the job, journals supported entirely by drug companies' revenues and controlled by the AMA, refused to publish any paper by anyone on Rife's therapy. Entire generations of medical students have graduated without hearing about Royal Rife's miracle therapy against numerous diseases.

The magnitude of such a crime eclipses mass murder in history. In 1960, the casualties from this disease exceeded the carnage of all wars America ever fought. In 1989, it was estimated that 40% of us will experience cancer at some time in our lifetime, and this number is growing. To this carnage you can add other poisons in our air, water and food that the FDA and EPA refuse to acknowledge as to their toxic effect. In Rife's lifetime, he witnessed progress of civilization from horse and buggy to jet planes. In the same time frame, he witnessed the epidemic of cancer increase from one in twenty-four Americans in 1905 to one in three in 1971, when Rife passed away.

It has reemerged in the underground medical/alternative health world. Fortunately, a few humanitarian doctors and engineers reconstructed his frequency instruments and kept Rife's genius alive. There is a wide variation in the cost and design and quality of the modern portable Rife frequency research instruments available. Costs can vary from $1,200 to $3,600. With the price there is no legitimate indicator of the technical competence in the design and performance of the instrument.

http://www.healingtools.tripod.com/rifestory.html

http://www.Pronto.com/Rife

http://www.resonantlight.com

It is a shame that some of these so-called friendly cancer centers do not evaluate some of these instruments. Big Pharma does not allow it since what can they do with an overabundance of overpriced chemotherapy drugs that kill? I lost a good friend to chemotherapy although after the third treatment, there was no indication of cancer cells in his body. Yet the doctors insisted on the full five-phase treatment.

One day, the name of Royal Raymond Rife may be placed in its rightful place as the giant of modern medical science. Until such time, his space age medical technology remains a secret except for the very few who seek it out.

http://medicaltruth.com/rife/sightings.html

One very interesting occurrence in my lifetime is this. Did you know that our President Ronald Reagan had cancer? He did not trust any cancer centers in this country or any oncologist. **He went to Germany to *get cured of cancer using oxygen treatment*.** Yes, oxygen, the element that sustains our life. It has been verified that the cancer virus cannot exist in an oxygenated system. Oxygen kills cancer cells as well as other pathogens. President Reagan lived for a good number of years, cancer-free, and passed away due to old age.

There are other numerous *natural* cures for cancer that are being suppressed by AMA, *the American Morbidity Association*, or the FDA, the Food and Drug Administration, or *Federal Death Agency*. All you have to do is access the information on the Internet to learn the truth. What AMA did to Royal Rife was the beginning of how well the American public can be controlled via government agencies using totalitarian, tyrannical control and silence others through intimidation, character assassination or even death.

What is really a crime against civilization is that we lost decades of additional fruitful testing against all different types of deadly microbes. There would not be a need for all the expensive treatments or poisonous inoculations since the only treatment would be sitting for a few minutes in front of a resonating frequency wavelength instrument, which would be doing its lethal work against the microbial disease. Some of the reports indicate that the treatment has to be limited in time since the circulatory and lymphatic system would be overloaded with dead organism body parts. How about devising such lethal frequency waves against the pests in agriculture, in place of the toxic chemicals?

Other reports that come to mind is the ferocious attack on any reports that come forward regarding the effectiveness of any natural solutions against numerous diseases. AMA, FDA, and Big Pharma, through their controlled news media, go into a frenzy, using the corrupt media to discredit any natural curative solutions. They just want to keep feeding us the health sick system that uses synthetic poisons. The other problem is that there are no requests for independent studies to confirm Big Pharma's claim as to the effectiveness to their synthetic medicines. When you get a prescription filled for one of the pharmaceutical medicines by the mainstream physicians, one also gets a list of about two pages of adverse side effects. Then you go back to get rid of your new complaint to get another prescription to eliminate your new derogatory side effect. Problem is, these medications do not cure the disease but alleviate the symptoms only. This is a managed health care system, as stated previously, to ensure wealth transfer from the patient to the health providers and their corrupt industry. Now you know why the health insurance rates are so expensive. What about the root cause of the problem? The medical schools do not provide any lessons regarding natural solutions. This would be contrary to the mainstream medical professions tyrannical philosophy as well as the international pharmaceutical monopoly. This is just one phase of destruction of America and also the civilization. Mass murder through medications and vaccinations. This can be interpreted as managing disease to do its lethal harm.

The other methods for the destruction of America were covered under the individual different headings initially. This is only a tip of the iceberg. There probably are other toxic chemicals that are probably proposed or in some form in planning stages. One has to notice how unknown organizations are spending thousands and millions of dollars to keep fluoridation of water plants, whenever it comes up to be continued or not, by municipalities. Prime example—City of Portland, Oregon. Extensive sums of money was spend to convince the people regarding continued fluoridation, or one can say continuing poisoning and dumbing the people due to the fluoride in their water. That was used extensively in Nazi Germany. The lie was perpetrated way back in the early fifties, ignoring scientific studies proving its lethal effect on health. The same can be said about aspartame. The data submitted to then more honest FDA officials was misleading and was not approved. But through some dishonest manipulations, it was finally approved, although studies indicated that it can cause tumors, but that information was conveniently withheld in their application. The Agricultural Association is pushing their agenda to add the toxin aspartame to increased dairy product, not being satisfied that it already is added to about five to six thousand different food products in one form or another. How about the American Grocery Association fighting and suing the State of Vermont regarding their law to label GMO products? The sad part is, the lawsuit is supported by the Big Food giants like Kellogg, Coca-Cola, Pepsi-Cola, General Mills, Campbell's, Kraft, and others. Most European countries, China, Russia, and now Canada are against GMO crops. How about the Indian farmers who had massive crop failure and committed suicide due to placing them in bankruptcy because of their GMO seeds that did not yield promised results. All other countries outside of the United States require extensive studies to prove their safety, but the dumbed-down American citizen has to accept whatever Monsanto tells them to accept, with the blessing of our FDA, EPA, etc., or whatever other federal agency. The question to be answered by FDA is, if the GMO food is so good for us, **then why is it banned by the White House and banned in Monsanto's cafeterias in their plants?**

You must understand one thing, dear reader—alternative medicine has a thousand-year history regarding its curative powers. The only difference now is that with our scientific analytical capabilities, we are able to identify the health improving ingredients in natural plants, herbs, natural oils, and/or combination of any of these including certain light beams. **Like Hippocrates said, let your food be your medicine, and the medicine be your food.** Yet the mainstream media and the pseudomedical profession regards these alternative doctors/healers as quacks. I believe it is the other way around—the misinformation is taught in our medical schools.

There are other inventions that have been suppressed by some powerful individuals or corporations which have robbed the American consumer of good environmentally safe inventions. One that comes to mind is the Nikola Tesla free electricity generator, using cosmic rays to generate electricity. This was silenced by withdrawing funds from him by J. P. Morgan since he could not gain any financial profit. Think about this—our electrical grid is in danger that can come from a terrorist attack, or equipment breakdown, which has happened. Using Tesla's free electrical generator, each home, business, building, or hospital would have an independent source of energy. It would also eliminate pollution since we would not need large amounts of fossil fuel usage to generate our energy. It is a completely free energy source and clean at that. Another source is the cold fusion. What happened to it? Another project silenced by Big Oil? How about using water as fuel in our cars, other engines or as stationary power source. It is used in the welding industry. Water is a perfect renewable energy source, and plentiful at that. Water, under an electric current, is broken into two elements—hydrogen and oxygen. When it is burned, there is a tremendous amount of energy released, and the nice part is, the byproduct is water. Systems do exist that can be applied to the automotive industry, but the Big Oil or government will not allow it. Some of the initial benefits reported regarding this added system to present engines is reduced pollutants into the air, improved fuel economy, cleaner burning engines, and less oil changes. For more information, please go to http://www.WantToKnow.com, access their index page, and select energy. This website belongs to PEERS, who have a wealth of information on numerous subjects of interest. Theirs is a compilation of news articles that in most instances have been suppressed by mainstream media. They have topics on vaccinations, GMO food, chemtrails, etc.

The use of lethal pesticides, insecticides, and herbicides has increased exponentially since the ban of DDT and other organochloride compounds that are lethal to our health. No one seems to care regarding the toxicity these compounds exert on us and our ecology. One way for the agricultural industry to control the pests would be to apply the Royal Rife principle that every living organism has a specific resonating frequency. Such a system would be highly advantageous for a number of reasons. First, it is lethal to target pests. Second, the killed pest is part of food chain and if consumed does not poison the consuming, bird, animal, or fish. Third, it does not harm the ecology, just to name a few advantages of such pest control system. It is also safe to human or animal life. Please understand that this is just my opinion. You must make up your own mind. All I am trying to do is steer you into a direction where you can access the above information to fabricate your own conclusion. My opinion is, we have to eliminate all these toxins to get to a better, healthy lifestyle. None of the so-called scientific studies compare healthy, free-from-toxin food, to the present toxin-infested processed food. I guess that some of these corporations in agriculture or medicine are afraid of what that kind of study would indicate. It probably would eliminate a number of health concerns that are present presently in our corrupt lifestyle. It would also decrease their inflated income. Too bad that our elected officials cater to these corporations instead being the watchdog of the constituents that elect them.

The first title of this book I used was *Silent Killers*. Reflecting on the massive information regarding these silent chemical toxins used, *Chemical Warfare on America* is a much better, appropriate title. Is it serving some unknown enemy's goal at present? Poisoning the world with chemtrails, aspartame, pesticides, herbicides and insecticides, and mass vaccinations must be a goal of an unknown enemy. Could it be the globalists that want to decrease the world's population?

Appendix

http://www.holisticmed.com/aspartame/aspfaq.html-3k,http://
www.medicinenet.com,.http://www.gazette.net
http://www.independent.co.uk
http://www.ncbi.nlm.nih.gov.http://www.opednews.com
http://www.yuobserver.com
http://www.zcommunications.org
http://www.smh.com.au
http://www.rense.com
http://www.flex-news-food.com
http://www.nzherald.co.nz
http://www.merrittnews.net
http://www.food.gov.uk
http://www.ncbi.nlm.nih.gov.
http://www.ncbi.nlm.nih.gov
http://www.huffingtonpost.com
http://www.latimes.com
http://www.ncbi.nlm.nih.gov
http://www.webmd.com
http://www.ncbi.nlm.nih.gov
http://www.holisticmed.com
http://www.mpwhi.com
.http://www.scoop.co.nz
http://www.neoucom.edu
http://www.medicinenet.com.
http:// www.ncbi.nlm.nih.gov
http://www.dailymail.co.uk
http://www.drbriffa.com

http://www.ncbi.nlm.nih.gov
http://www.eastbourneherald.co.uk
http://www.imdb.com.

To read the prescribing information for each individual vaccine, see List of Licensed Vaccines
Vaccine types, Vaccination schedule, Adjuvant, Preservative, Cell culture, Growth medium

References

The initial list is based on information from the Centers for Disease Control and Prevention (CDC) and the Food and Drug Administration (FDA) and thus limited to US-approved vaccines.

Background History of Author

I was born in Poland in 1935, in a little town called Boryslaw, which, after the war became part of Ukraine, as part of Eastern Poland, was annexed by USSR. There is little that comes to my memory from those first five years. I remember that we lived in a town called Bochnia. When I was enrolled in the first grade at age six, it was the first time that I had to learn two languages in school—Polish and German. Whatever memories I have from 1941 and onward were usually what a young boy at that age was interested in, namely sports and occasional trips to my grandparents, who had a farm a few kilometers from the town. There is one incident that stayed in my mind till today. One day the adults were talking that two German soldiers were killed in the town where we lived. Then about two days later, I could hear machine gun fire. I thought that it must have been some practice by the German troops. I heard the adults saying that that day, the German soldiers cordoned off the town square, picked out about a hundred fifty men, marched them out into nearby woods, and machine-gunned them down. To this day I can still recollect the sound of those machine guns. I do remember that my dad was the town electrician. In 1944, in August, my aunt Mali came to visit us in Poland. She lived in Austria. She married an Austrian citizen in 1933. I remember I came home in the early afternoon, and my mother told me to get ready because we were leaving the country and going to Austria with our aunt. Little did I understand that this was due to the war. My dad did not want the family to stay in Poland under the communist Russian rule. From August 1944 to May 1945 we stayed with my aunt in a town called Hallein, just outside of Salzburg, Austria. I was also enrolled in the Austrian schools. Since I was eight years old, I automatically became a member of the youth club called Hitlerjugend.

All boys at that age had to be in that group. During the air raids, the town was almost 60% destroyed. I remember running to the tunnel that was provided by the town, whenever an air raid signal was given. The mountain shook as the bombs were exploding outside. After one of those air raids, I remember seeing my friend in a coffin, a day after we played together.

When the end of the war came in May 1945, refugee camps were organized for different nationalities. These consisted of the people displaced from their native countries and brought to Germany and Austria, to work in their factories or fields. We joined the Polish camp that was just outside of Salzburg, close to the white castle that stands on a hill above Salzburg. I remember my dad taking me on a visit to that castle. I remember seeing the white castle of Salzburg every day in the background. The camp that was assigned to Polish refugees was a German army camp during the war.

Within a few months, transports were organized for people who wanted to return to their native land. Transports were also going to Poland. One of my dad's friends went on the very first group to return to Poland. The transports were scheduled about once every three months. My dad's friend smuggled a letter back to us, stating that this was not the communist philosophy he studied. He mentioned that many of the returned Polish people were sent to camps. He advised us to not to come back to Poland.

My parents then decided not to return to Poland. Word came that people who did not want to return to their native lands could apply to immigrate to other countries. My father applied to go to the United States. Since his mother's brother immigrated to the United States in 1913, and we needed a sponsor to go to the United States. My dad wrote a letter to his uncle addressed in the following manner: John Woldrych, c/o Ford Motor Co., Detroit Michigan, USA. A conscientious worker delivered the letter to my dad's uncle.

My parents also learned that my mother's younger brother, Henry, escaped from Poland and joined the Polish Army as a combat engineer. We also learned that my mother's brother was a member of the Polish Second Corps, attached to the British military, and was stationed presently in Ancona, Italy. In August 1946 my family decided to join my mother's brother in Italy, so we left the camp, traveling by train to Innsbruck, Austria. There we were met by some Polish organizers to smuggle us into Italy. We crossed the border at night, on foot above the border railroad station at Brenner Pass. On the Italian side, trucks were waiting for us to take us to Ancona. When we arrived there, we learned that my mother's brother, my uncle, was transported to Great Britain. In December 1946, a transport was organized to take all the Polish refugees at the Ancona camp to Great Britain. We were put on a train in Italy that traveled through parts of Austria, Germany, and France, to the French port of Calais. After a night there, we were crammed onto a ship called *Columbia*, to cross the English Channel. I remember most of the people were seasick for the duration of the crossing. On December 6, 1946, we landed in England. On the British side, we were transported to a Polish Camp in Pulborough, Sussex, England. There we met up with my mother's brother, Henry.

While in this camp, we were contacted by UN authorities, who were looking for us regarding our emigration status to the United States. We also received a letter from my uncle in the States. A conscientious worker at Ford Motor Co. located my uncle and delivered the letter to him. We learned that if we had stayed in Austria, we would have been in the States in 1948. From the Pulborough camp, we were transferred to another camp just outside of Cirencester, England. During Roman times, it was the second largest Roman city in England.

What was always organized were schools for the young people in each camp. Having located a city dump, which contained numerous thrown out bicycles, I was able with someone's help to put a bicycle together. I finished the camp grade school and enrolled myself into the British school system, which accelerated my learning of the English language. The bicycle came in quite handy since I had to pedal about three miles to the school in Cirencester. I spent a year and a half at the Cirencester Secondary Modern School (that was the name). On graduation, the teachers at that school enrolled me into so-called Stroud Technical College. I believe that is equivalent to a middle school here in the States.

The town of Stroud was located quite a few miles from where I lived. My daily travel to school consisted of the following. I had to ride my bicycle to the railroad station in Cirencester, which was about three miles from the camp. There I would take a train to another station since the city of Cirencester was just outside of the main railway line. So I had to take connecting spur to the main railway line, where I would catch another train to get to Stroud. The school system provided me with necessary seasonal railroad ticket. I left the camp at 7:00 a.m. pedaled to the station and took a train to another station, and from there on another train to Stroud, where I walked to the school and had to be in class at 9:00 a.m. School lasted till 3:00 p.m. Travel consumed two hours to school and two hours back home. I had about thirteen different subjects. Some subjects were twice a week and some were three times a week. Needless to say, I was challenged quite a bit.

In late 1950, we were finally called to the US embassy in London for our immigration to the United States. In March 1951, we were notified that we were scheduled to immigrate to the United States, and we received sea passage paid for by the British government. We were given passage on HMS *Scythia*, a Cunard Shipping Line ship. We left Port of Liverpool in early March. On March 21, 1951, we landed in port of New York City. After clearing customs and obtaining railroad tickets, we were finally on our way to Detroit, Michigan. My dad's relatives met us at the grand station in Detroit and took us to their home.

We stayed with our relatives for about two to three weeks, after which they found us a rental home. They paid the first month's rent and food. We were on our own after that. My sister hired as a full-time babysitter to keep us going. My dad, in order to find employment, had to submit so-called first papers, meaning that we intended to become US citizens after the five-year waiting period. My dad finally found work as an electromechanic repairing machinery. He translated his professional papers and was hired as an electrician by General Motors, where he worked for over twenty-two years. I remember when my dad was earning $60 per week. It was sufficient to pay for our rent and utilities, and we would spend like $20 a week for groceries. For that amount, we would get six large bags full of groceries.

I was enrolled in a high school, where I could only take four to five subjects a semester—a far cry from what I had to study in England. I could do my homework during study periods. This left me a lot of free time. I started going to the library and took out books regarding American history and quite a few novels regarding the American Revolution. I graduated from Chadsey High School in June 1954. I received two scholarships—one from Wayne State University and one from the University of Michigan—that covered my tuition expenses. In 1954, the tuition for about fifteen credit hours at University of Michigan was $100 per semester, and it included tickets to the football games. The tuition at Wayne State University was $110 per semester. I graduated with a magna cum laude achievement from my high school. In the fall, I started my college career at the University of Michigan.

In the fall of 1956, I transferred to Wayne State University and decided to get a degree in analytical chemistry. I enjoyed college life too much and when I messed up one of my semesters, I was drafted into the military in July 1959. I spent my basic training at Fort Leonard Wood, in Missouri. It was also called little Korea. After basic training, I was sent to a quartermaster school at Fort Lee, Virginia. In early December 1959, I received orders that sent me to Korea, to work in the Army Petroleum Depot at Inchon, Korea. I spent eighteen months in Korea and upon return to the States, I was discharged at Oakland Army Terminal, California, and given a one-way airline ticket home. This was in May 1961.

In the fall, I returned to Wayne State University to finish my studies and graduated with a bachelor of science degree in analytical chemistry. In the fall, I started to work for Federal Food and Drug Administration as a bench chemist, analyzing different drugs and food contents for proper amount. The starting salary was $600 a month. I worked for FDA till February 1966. I had two analytical papers published while working for FDA. These were "Quantitative Analysis of Acetylcarbromal" in 1963 and "Separation of Dextromethorphan Hydro-Bromide and Glyceryl Guaiacolate Ether in Elixirs" in 1964. These are a combination of drugs sold as antihistamines.

I started working for Chrysler Corporation, earning about $650 a month in their analytical chemical laboratory. Three months later, I was transferred to a new group that was involved with analysis of vehicle emissions. I spent the next seventeen years in that group. I saw it expand from initial six people to about fifty. That is just to operate different instruments in the measurement and analysis of emissions from the vehicles. One of my more memorable assignments was when my supervisor asked me to go on second shift to resolve labor problems. There were a lot of union grievances written as well as foreman's reports against mechanics. My first day on second shift, the second shift steward approached me and said, "My name is Joe Mundell, I am the second shift steward, and I am known to be a hard nose."

My answer was, "It is nice to meet you Joe. As far as what kind of a person you are, I did not have the pleasure to work with you."

Joe was one of the old-timers who remembered the start of the unions in the auto industry. I found him to be a very knowledgeable mechanic, from whom I learned quite a bit about cars and how to fix them. I also enjoyed hearing from Joe about the authoritarian conditions that the mechanics and also engineers had to endure prior to unionization of the auto industry. Joe and I came to an agreement right after our first meeting that before any grievances or foreman reports were written, we would discuss the problem among ourselves, thereby resolving any problems. Needless to say, the human resources department was wondering what happened on the second shift since the grievances and foreman reports ceased. One of my other assignments was also to implement and conduct instrumental analysis course for vehicle emissions, designed for mechanics and engineers. I had two internal publications: *Reference Manual on Emission Instrumentation* and *Emission Test Procedures.*

After seventeen years, I transferred to a quality group, following warranty claims against engines and transmissions. This also gave me an opportunity to travel and do dealer visits to follow up on problems reported by the dealers on engines and transmissions. I retired in 1999 but joined an interior trim department with Chrysler Corporation on a per diem basis. When I started working for FDA, I was earning $600 a month. When I retired from Chrysler, I was earning $6,500 a month. Over the years, due to union contracts, I saw my earnings go up. I remember talking with a day shift steward regarding upcoming contract talks. He told me about what the union was going to demand in oncoming contract talks, which was to include annual improvement factors of about 2 to 3% per year, plus continuing the cost of living factor. I remember telling him that that would price us out of the world market and contribute to inflation. His reply was that the big bosses give themselves large bonuses. I told him then to go after profit sharing. These annual improvement raises and cost of living factors contributed to the inflation in the country.

While working for the interior trim department, I was involved in the Chrysler minivan platform. I spent a lot of time in those respective assembly plants where the minivan was built, which was predominantly in Windsor, Ontario, Canada, St. Louis, Missouri when it was still open and also numerous trips to Graz, Austria, where Chrysler was building the minivan for European and Middle Eastern markets.

While I was still working, I became quite interested in alternative medicine. My mother made an ointment that was very effective against bruises. I started to sign up to some of the alternative medicine reports. One of the early reports was by Dr. David Williams, who publishes a monthly newsletter titled "Alternatives for the

Health Conscious Individual." As luck would have it, Dr. Williams wrote about one of the ingredients within the first four months after I subscribed to it. In another three months, he wrote about the second ingredient that I started to use in my formula in this all natural ointment.

While I was still working, I decided to have some of my friends and fellow coworkers try it out. I received numerous confirmations regarding the effectiveness of the ointment, as it was listed by Dr. Williams. I compiled testimonials regarding the usage of the ointment on numerous different applications by my coworkers, family, and friends. I had to separate the testimonials into three distinct categories: sprains, strains, and bruises; insect bites of different origin; and skin rashes, minor burns, and slight wounds.

If one would like to obtain further information regarding this ointment, please send a stamped, self addressed envelope to ER, P. O. Box 961764, El Paso, Texas, U.S.A.

CPSIA information can be obtained
at www.ICGtesting.com
Printed in the USA
FSHW02n1559140818
51276FS